Kubernetes Security and Observability

A Holistic Approach to Securing Containers and Cloud Native Applications

Brendan Creane and Amit Gupta

Beijing • Boston • Farnham • Sebastopol • Tokyo

Kubernetes Security and Observability

by Brendan Creane and Amit Gupta

Published by O'Reilly Media, Inc., 1005 Gravenstein Highway North, Sebastopol, CA 95472.

O'Reilly books may be purchased for educational, business, or sales promotional use. Online editions are also available for most titles (*http://oreilly.com*). For more information, contact our corporate/institutional sales department: 800-998-9938 or *corporate@oreilly.com*.

Acquisitions Editor: John Devins
Development Editor: Virginia Wilson
Production Editor: Beth Kelly
Copyeditor: J. M. Olejarz
Proofreader: Kim Wimpsett
Indexer: Sue Klefstad
Interior Designer: David Futato
Cover Designer: Karen Montgomery
Illustrator: Kate Dullea

November 2021: First Edition

Revision History for the First Edition

2021-11-26: First Release

See *http://oreilly.com/catalog/errata.csp?isbn=9781098107109* for release details.

This work is part of a collaboration between O'Reilly and Tigera. See our statement of editorial independence (*https://oreil.ly/editorial-independence*).

978-1-098-10710-9

[LSI]

Table of Contents

Preface

Kubernetes is not secure by default. Existing approaches to enterprise and cloud security are challenged by the dynamic nature of Kubernetes and the goal of increased organizational agility that often comes with using it. Successfully securing, observing, and troubleshooting mission-critical microservices in this new environment requires a holistic understanding of a breadth of considerations. These include organizational challenges, how new cloud native approaches can help meet the challenges, and the new best practices and how to operationalize them.

While there is no shortage of resources on Kubernetes, navigating through them and formulating a comprehensive security and observability strategy can be a daunting task and in many cases leads to gaps that significantly undermine the desired security posture.

That's why we wrote this book—to guide you toward a holistic security and observability strategy across the breadth of these considerations and to give you best practices and tools to help you as you move applications to Kubernetes.

Over our years of working at Tigera and building Calico, a networking and security tool for Kubernetes, we have gotten to see the user journey up close. We have seen many users focus on getting their workloads deployed in Kubernetes without thinking through their security or observability strategy, and then struggle as they try to understand how to secure and observe such a complex distributed system. Our goal with this book is to help minimize this pain as much as possible by sharing with you what we've learned. We mention a number of tool examples throughout, and Calico is among them. We believe that Calico is an excellent and popular option, but there are many good tools, like Weave Net, VMware Tanzu, Aqua Security, and Datadog, to choose from. Ultimately, only you can decide which is best for your needs.

The Stages of Kubernetes Adoption

Any successful Kubernetes adoption journey follows three distinct stages:

The learning stage
: As a new user, you begin by learning how Kubernetes works, setting up a sandbox environment, and starting to think about how you can use Kubernetes in your environment. In this stage you want to leverage the online Kubernetes resources available and use open source technologies.

The pilot/pre-production stage
: Once you familiarize yourself with Kubernetes and understand how it works, you start thinking about a high-level strategy to adopt Kubernetes. In this stage you typically do a pilot project to set up your cluster and onboard a couple of applications. As you progress in this stage, you will have an idea about which platforms you're going to use and whether they will be on-premise or in the cloud. If you choose cloud, you will decide whether to host the cluster yourself or leverage a managed Kubernetes service from a cloud provider. You also need to think about strategies to secure your applications. By this time, you would have realized that Kubernetes is different due to its declarative nature. This means that the platform abstracts a lot of details about the network, infrastructure, host, etc., and therefore makes it very easy for you to use the platform for your applications. Because of this, the current methods you use to secure your applications, infrastructure, and networks simply do not work, so you now need to think about security that is native to Kubernetes.

The production stage
: By this point, you have completed your pilot project and successfully onboarded a few applications. Your focus is on running mission-critical applications in production and on considering whether to migrate most of your applications to Kubernetes. In this stage you need to have detailed plans for security, compliance, troubleshooting, and observability in order to safely and efficiently move your applications to production and realize all the benefits of the Kubernetes platform.

The popularity and success of Kubernetes as a platform for container-based applications has many people eager to adopt it. In the past couple of years, there has been an effort by managed Kubernetes service providers to innovate and make adoption easier. New users may be tempted to go past the learning and pilot stages in order to get to the production stage quickly. We caution against skipping due diligence. You must consider security and observability as critical first steps before you onboard mission-critical applications to Kubernetes; your Kubernetes adoption is incomplete and potentially insecure without them.

Who This Book Is For

This book is for a broad range of Kubernetes practitioners who are in the pilot/preproduction stage of adoption. You may be a platform engineer or part of the security or DevOps team. Some of you are the first in your organization to adopt Kubernetes and want to do security and observability right from the start. Others are helping to establish best practices within an organization that has already adopted Kubernetes but has not yet solved the security and observability challenges Kubernetes presents. We assume you have basic knowledge of Kubernetes—what it is and how to use it as an orchestration tool for hosting applications. We also assume you understand how applications are deployed and their distributed nature in a Kubernetes cluster.

Within this broad audience, there are many different roles. Here is a nonexhaustive list of teams that help design and implement Kubernetes-based architectures that will find value in this book. Please note that the role names may be different in your organization, so please look at the responsibilities for each to identify the corresponding role in your organization. We will use these names throughout the book to help you understand how a concept impacts each role.

The Platform Team

The platform engineering team is responsible for the design and implementation of the Kubernetes platform. Many enterprises choose to implement a container as a service platform (CaaS) strategy. This is a platform that is used across the enterprise to implement container-based workloads. The platform engineering team is responsible for the platform components and provides them as a service to application teams. This book helps you understand the importance of securing the platform and best practices to help secure the platform layer—that way you can provide application teams a way to onboard applications on a secure Kubernetes platform. It will also help you learn how to manage the security risk of new applications to the platform.

The Networking Team

The networking team is responsible for integrating Kubernetes clusters in an enterprise network. We see these teams play different roles in an on-premise deployment of Kubernetes and in a cloud environment where Kubernetes clusters are self-hosted or leverage a managed Kubernetes service. You will understand the importance of network security and how to build networks with a strong security posture. Best practices for exposing applications outside the Kubernetes platform as well as network access for applications to external networks are examples of topics covered in this book. You will also learn how to collaborate with other teams to implement network security to protect elements external to Kubernetes from workloads inside Kubernetes.

The Security Team

The security team in enterprises is the most impacted by the movement toward cloud native applications. Cloud native applications are those designed for cloud environments and are different from traditional applications. As an example, these applications are distributed across the infrastructure in your network. This book will help you understand details about how to secure a Kubernetes platform that is used to host applications. It will provide you a complete view of how to secure mission-critical workloads. You will learn how to collaborate with various teams to effectively implement security in the new and different world of Kubernetes.

The Compliance Team

The compliance team in an enterprise is responsible for ensuring operations and processes in an organization to meet the requirements of compliance standards adopted by an organization. You will understand how to implement various compliance requirements and how to monitor ongoing compliance in a Kubernetes-based platform. Note that we will not cover detailed compliance requirements and various standards, but we will provide you with strategies, examples, and tools to help you meet compliance requirements.

The Operations Team

The operations team is the team of developers/tools/operations engineers responsible for building and maintaining applications. They are also known as DevOps or site reliability engineers (SREs). They ensure that applications are onboarded and meet the required service level agreements (SLAs). In this book you will learn about your role in securing the Kubernetes cluster and collaboration with the security team. We will cover the concept of shift-left security, which says security needs to happen very early in the application development life cycle. Observability in a Kubernetes platform means the ability to infer details about the operation of your cluster by viewing data from the platform. This is the modern way of monitoring a distributed application, and you will learn how to implement observability and what its importance to security is.

What You Will Learn

In this book you will learn how to think about security as you implement your Kubernetes strategy, from building applications to building infrastructure to hosting applications to deploying applications to running applications. We will present security best practices for each of these with examples and tools to help you secure your Kubernetes platform. We will cover how to implement auditing, compliance, and other enterprise security controls like encryption.

You will also learn best practices with tools and examples that show you how to implement observability and demonstrate its relevance to security and troubleshooting. This enhanced visibility into your Kubernetes platform will drive actionable insights relevant to your unique situation.

By the end of the book, you will be able to implement these best practices for security and observability for your Kubernetes clusters.

Conventions Used in This Book

The following typographical conventions are used in this book:

Italic
: Indicates new terms, URLs, email addresses, filenames, and file extensions.

`Constant width`
: Used for program listings, as well as within paragraphs to refer to program elements such as variable or function names, databases, data types, environment variables, statements, and keywords.

`Constant width bold`
: Shows commands or other text that should be typed literally by the user.

`Constant width italic`
: Shows text that should be replaced with user-supplied values or by values determined by context.

This element signifies a general note.

Using Code Examples

Supplemental material (code examples, exercises, etc.) is available for download at *https://github.com/tigera/k8s-security-observability-book*.

If you have a technical question or a problem using the code examples, please send email to *bookquestions@oreilly.com*.

This book is here to help you get your job done. In general, if example code is offered with this book, you may use it in your programs and documentation. You do not need to contact us for permission unless you're reproducing a significant portion of the code. For example, writing a program that uses several chunks of code from this book does not require permission. Selling or distributing examples from O'Reilly books does require permission. Answering a question by citing this book and quoting example code does not require permission. Incorporating a significant amount of example code from this book into your product's documentation does require permission.

We appreciate, but generally do not require, attribution. An attribution usually includes the title, author, publisher, and ISBN. For example: "*Kubernetes Security and Observability* by Brendan Creane and Amit Gupta (O'Reilly). Copyright 2022 O'Reilly Media, 978-1-098-10710-9."

If you feel your use of code examples falls outside fair use or the permission given above, feel free to contact us at *permissions@oreilly.com*.

O'Reilly Online Learning

For more than 40 years, *O'Reilly Media* has provided technology and business training, knowledge, and insight to help companies succeed.

Our unique network of experts and innovators share their knowledge and expertise through books, articles, and our online learning platform. O'Reilly's online learning platform gives you on-demand access to live training courses, in-depth learning paths, interactive coding environments, and a vast collection of text and video from O'Reilly and 200+ other publishers. For more information, visit *http://oreilly.com*.

How to Contact Us

Please address comments and questions concerning this book to the publisher:

O'Reilly Media, Inc.
1005 Gravenstein Highway North
Sebastopol, CA 95472
800-998-9938 (in the United States or Canada)
707-829-0515 (international or local)
707-829-0104 (fax)

We have a web page for this book, where we list errata, examples, and any additional information. You can access this page at *https://oreil.ly/KSO*.

Email *bookquestions@oreilly.com* to comment or ask technical questions about this book.

For news and information about our books and courses, visit *http://oreilly.com*.

Find us on Facebook: *http://facebook.com/oreilly*

Follow us on Twitter: *http://twitter.com/oreillymedia*

Watch us on YouTube: *http://youtube.com/oreillymedia*

Acknowledgments

It was a great experience writing this book and it would not have been possible without the help, support and guidance of several people. Firstly, we want to thank the community, developers and maintainers of Project Calico, your innovation and contributions to Kubernetes and Kubernetes security and observability, have enabled us to write this book. The amazing engineering and the security research teams at Tigera have built products to address the complex challenges for security and observability, and this enabled us to get a clear understanding of the challenges facing the users. This was very helpful as we wrote this book to guide users to a holistic security and observability solution.

We also wanted to thank the reviewers who provided their opinions and subject matter expertise. Their comments and guidance have greatly enriched the content of this book. Special mention to Manish Sampat, Alex Pollitt, Virginia Wilson, Seth Vargo, Tim Mackey, Ian Lewis, Puja Absassi, and Jose Ruiz—you are awesome!

Finally, we want to thank everyone in the community that is contributing to Kubernetes security and observability. It is amazing to see the innovation in this area, and we are thrilled to be involved with Kubernetes security and observability.

CHAPTER 1

Security and Observability Strategy

In this chapter, we will cover a high-level overview of how you can build a security and observability strategy for your Kubernetes implementation. Subsequent chapters will cover each of these concepts in more detail. You need to think about a security strategy when you are in the pilot/pre-production phase of your Kubernetes journey, so if you are part of the security team, this chapter is very important. If you are part of the network, platform, or application team, this chapter shows how you can be a part of the security strategy and discuss the importance of collaboration between the security, platform, and application teams.

We will cover the following concepts that will guide you with your security and observability strategy:

- How securing Kubernetes is different from traditional security methods
- The life cycle of deploying applications (workloads) in a Kubernetes cluster and best practices for each stage
- How you should implement observability to help with security
- Well-known security frameworks and how to use them in your security strategy

Security for Kubernetes: A New and Different World

In this section we'll highlight how Kubernetes is different and why traditional security methods do not work in a Kubernetes implementation.

As workloads move to the cloud, Kubernetes is the most common orchestrator for managing them. The reason Kubernetes is popular is its declarative nature: It abstracts infrastructure details and allows users to specify the workloads they want to run and the desired outcomes. The application team does not need to worry about how workloads are deployed, where workloads are run, or other details like networking; they just need to set up configurations in Kubernetes to deploy their applications.

Kubernetes achieves this abstraction by managing workload creation, shutdown, and restart. In a typical implementation, a workload can be scheduled on any available resource in a network (physical host or virtual machine) based on the workload's requirements. A group of resources that a workload runs on is known as a *Kubernetes cluster*. Kubernetes monitors the status of workloads (which are deployed as pods in Kubernetes) and takes corrective action as needed (e.g., restarting unresponsive nodes). It also manages all networking necessary for pods and hosts to communicate with each other. You have the option to decide on the networking technology by selecting from a set of supported network plug-ins. While there are some configuration options for the network plug-in, you will not be able to directly control networking behavior (either for IP address assignment or in typical configurations where the node is scheduled).

Kubernetes is a different world for security teams. Their traditional method would be to build a "network of machines" and then onboard workloads (applications). As a part of onboarding, the process was to assign IPs, update networking as needed, and define and implement network access control rules. After these steps, the application was ready for users. This process ensured that security teams had a lot of control and could onboard and secure applications with ease. The applications were easy to secure, as applications were static in terms of assigned IPs, where they were deployed, etc.

In the Kubernetes world, workloads are built as container images and are deployed in a Kubernetes cluster using a configuration file (yaml). This is typically integrated in the development process, and most development teams use continuous integration (CI) and continuous delivery (CD) to ensure speedy and reliable delivery of software. What this means is that the security team has limited visibility into the impact of each application change on the security of the cluster. Adding a security-review step to this process is counterproductive, as the only logical place to add that is when the code is being committed. The development process after that point is automated, and disrupting it would conflict with the CI/CD model. So how can you secure workloads in this environment?

In order to understand how to secure workloads in Kubernetes, it is important to understand the various stages that are part of deploying a workload.

Deploying a Workload in Kubernetes: Security at Each Stage

In the previous section, we described the challenge of securing applications that are deployed using the CI/CD pipeline. This section describes the life cycle of workload deployment in a Kubernetes cluster and explains how to secure each stage. The three stages of workload deployment are the build, deploy, and runtime stages. Unlike traditional client-server applications where an application existed on a server (or a cluster of servers), applications in a Kubernetes deployment are distributed, and the Kubernetes cluster network is used by applications as a part of normal operation. Here are a few things to consider because of this configuration:

- You need to consider security best practices as workloads and infrastructure are built. This is important due to the fact that applications in Kubernetes are deployed using the CI/CD pipeline.
- You need to consider security best practices when a Kubernetes cluster is deployed and applications are onboarded.
- Finally, applications use the infrastructure and the Kubernetes cluster network for normal operation, and you need to consider security best practices for application runtime.

Figure 1-1 illustrates the various stages and aspects to consider when securing workloads in a Kubernetes environment.

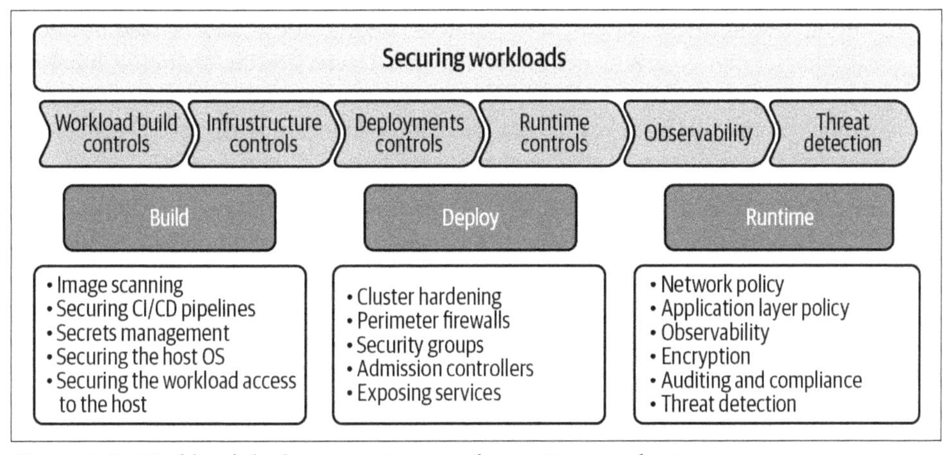

Figure 1-1. Workload deployment stages and security at each stage

The boxes below each stage describe various aspects of security that you need to consider for that stage:

- The build stage is where you create (build) software for your workload (application) and build the infrastructure components (host or virtual machines) to host applications. This stage is part of the development cycle, and in most cases the development team is responsible for it. In this stage you consider security for the CI/CD pipeline, implement security for image repositories, scan images for vulnerabilities, and harden the host operating system. You need to ensure that you implement best practices to secure the image registry and avoid compromising the images in the image registry. This is generally implemented by securing access to the image registry, though a lot of users have private registries and do not allow images from public registries. Finally, you need to consider best practices for secrets management; secrets are like passwords that allow access to resources in your cluster. We will cover these topics in detail in Chapter 3. We recommend that when you consider security for this stage, you should collaborate with the security team so that security at this stage is aligned with your overall security strategy.
- The next stage, deploy, is where you set up the platform that runs your Kubernetes deployment and deploy workloads. In this stage you need to think about the security best practices for configuring your Kubernetes cluster and providing external access to applications running inside your Kubernetes cluster. You also need to consider security controls like policies to limit access to workloads (pod security policies), network policies to control applications' access to the platform components, and role-based access control (RBAC) for access to resources (for example, service creation, namespace creation, and adding/changing labels to pods). In most enterprises the platform team is responsible for this stage. As a member of the platform team, you need to collaborate with both the development and the security teams to implement your security strategy.
- The final stage is the runtime stage, where you have deployed your application and it is operational. In this stage you need to think about network security, which involves controls using network policy, threat defense (using techniques to detect and prevent malicious activity in the cluster), and enterprise security controls like compliance, auditing, and encryption. The security team is responsible for this stage of the deployment. As a member of the security team, you need to collaborate with the platform and development teams as you design and implement runtime security. Collaboration between teams (development, platform, and security) is very important for building an effective security strategy. We recommend that you ensure all these teams are aligned.

Note that unlike with traditional security strategies, where security is enforced at a vantage point (like the perimeter), in the case of a Kubernetes cluster, you need to implement security at each stage. In addition, all teams involved (application, platform, and security) play a very important role in implementing security, so the key to implementing a successful strategy is collaboration between teams. Remember, security is a shared responsibility. Let's explore each stage and the techniques you can use to build your strategy.

Build-Time Security: Shift Left

This section will guide you through various aspects of build-time security with examples.

Image scanning

During this stage, you need to ensure that applications do not have any major unpatched issues that are disclosed as common vulnerability enumerations (CVEs) in the National Vulnerability Database, and that the application code and dependencies are scanned for exploits and vulnerable code segments. The images that are built and delivered as containers are then scanned for unpatched critical or major vulnerabilities disclosed as CVEs. This is usually done by checking the base image and all its packages against a database that tracks vulnerable packages. In order to implement scanning, there are several tools, both open source and commercial, that are available to you. For example, Whitesource, Snyk, Trivy, Anchor, and even cloud providers like Google offer scanning of container images. We recommend that you select a scanning solution that understands how containers are built and scans not only the operating system on the host but also base images for containers. Given the dynamic nature of Kubernetes deployments, it is very important for you to secure the CI/CD pipeline; code and image scanning needs to be a part of the pipeline, and images being delivered from the image registry must be checked for compromise. You need to ensure access to the registry is controlled to avoid compromise. The popular term to describe this stage is *shifting security left toward the development team*, also known as *shift-left security*.

Host operating system hardening

Here you must ensure that the application being deployed is restricted to having the required privileges on the host where it is deployed. To achieve this, you should use a hardened host operating system that supports controls to enable restricting applications to only necessary privileges like system calls and file system access. This allows you to effectively mitigate attacks related to *privilege escalation*, where a vulnerability in the software being deployed in a container is used to gain access to the host operating system.

Minimizing the attack surface: Base container images

We recommend you review the composition of the container image and minimize software packages that make up the base image to include only packages that are absolutely necessary for your application to run. In Dockerfile-based container images, you can start with a parent image and then add your application to the image to create a container image. For example, you could start by building a base image in Docker using the `FROM scratch` directive, which will create a minimal image. You can then add your application and required packages, which will give you complete control of the composition of your container images and also help with CVE management, as you do not need to worry about patching CVEs in packages in a container image that aren't required by your application. In case building a scratch image is not a viable option, you can consider starting with a distroless image (a slimmed-down Linux distribution image) or an Alpine minimal image as the base images for your container.

These techniques will help you design and implement your build-time security strategy. As a part of the development team, you will be responsible for designing and implementing build-time security in collaboration with the platform and security teams to ensure it is aligned with the overall security strategy. We caution against believing the myth that shift-left security can be your whole security strategy. It is incorrect, and a naive approach to securing workloads. There are several other important aspects, such as deploy and runtime security, that need to be considered as part of your security strategy as well.

Deploy-Time Security

The next stage in securing workloads is to secure the deployment. To accomplish this, you have to harden your Kubernetes cluster where the workloads are deployed. You will need a detailed review of the Kubernetes cluster configuration to ensure that it is aligned with security best practices. Start by building a trust model for various components of your cluster. A trust model is a framework where you review a threat profile and define mechanisms to respond to it. You should leverage tools like role-based access control (RBAC), label taxonomies, label governance, and admission controls to design and implement the trust model. These are mechanisms to control access to resources and controls and validation applied at resource creation time. These topics are covered in detail in Chapters 3, 4, and 7. The other critical components in your cluster are the Kubernetes datastore and Kubernetes API server, and you need to pay close attention to details like access control and data security when you design the trust model for these components. We recommend you use strong credentials, public key infrastructure (PKI) for access, and transport layer security (TLS) for data in transit encryption. Securing the Kubernetes APT and the datastore is covered in detail in Chapter 2.

You should think of the Kubernetes cluster where mission-critical workloads are deployed as an entity and then design a trust model for the entity. This requires you to review security controls at the perimeter, which will be challenging due to the Kubernetes deployment architectures; we will cover this in the next section. For now, let's assume the current products that are deployed at the perimeter, like web access control gateways and next-generation firewalls, are not aware of Kubernetes architecture. We recommend you tackle this by building integrations with these devices, which will make them aware of the Kubernetes cluster context so they can be effective in applying security controls at the perimeter. This way you can create a very effective security strategy where the perimeter security devices work in conjunction with security implemented inside your Kubernetes cluster. As an example, say you need to make these devices aware of the identity of your workloads (IP address, TCP/UDP port, etc.). These devices can effectively protect the hosts that make up your Kubernetes cluster, but in most cases they cannot distinguish between workloads running on a single host. If you're running in a cloud provider environment, you can use security groups, which are virtual firewalls that allow access control to a group of nodes (such as EC2 instances in Amazon Web Services) that host workloads. Security groups are more aligned with the Kubernetes architecture than traditional firewalls and security gateways; however, even security groups are not aware of the context for workloads running inside the cluster.

To summarize, when you consider deploy-time security, you need to implement a trust model for your Kubernetes cluster and build an effective integration with perimeter security devices that protect your cluster.

Runtime Security

Now that you have a strategy in place to secure the build and deploy stages, you need to think about runtime security. The term *runtime security* is used for various aspects of securing a Kubernetes cluster, for example on a host running software, but any configuration that protects the host and workloads from unauthorized activity (e.g., system calls, file access) is also called runtime security. Chapter 4 will cover host and workload runtime security in detail. In this section we will focus on the security best practices needed to ensure the secure operation of the Kubernetes cluster network. Kubernetes is an orchestrator that deploys workloads and applications across a network of hosts. You must consider network security as a very important aspect of runtime security.

Kubernetes promises increased agility and the more efficient use of compute resources, compared with the static partitioning and provisioning of servers or VMs. It does this by dynamically scheduling workloads across the cluster, taking into account the resource usage on each node, and connecting workloads on a flat network. By default, when a new workload is deployed, the corresponding pod could be scheduled on any node in the cluster, with any IP address within the pod IP address.

If the pod is later rescheduled elsewhere, then it will normally get a different IP address. This means that pod IP addresses need to be treated as ephemeral. There is no long-term or special meaning associated with pod IP addresses or their location within the network.

Now consider traditional approaches to network security. Historically, in enterprise networks, network security was implemented using security appliances (or virtual versions of appliances) such as firewalls and routers. The rules enforced by these appliances were often based on a combination of the physical topology of the network and the allocation of specific IP address ranges to different classes of workloads.

As Kubernetes is based on a flat network, without any special meaning for pod IP addresses, very few of these traditional appliances are able to provide any meaningful workload-aware network security and instead have to treat the whole cluster as a single entity. In addition, in the case of east-west traffic between two pods hosted on the same node, the traffic does not even go via the underlying network. So these appliances won't see this traffic at all and are essentially limited to north-south security, which secures traffic entering the cluster from external sources and traffic originating inside the cluster headed to sources outside the cluster.

Given all of this, it should be clear that Kubernetes requires a new approach to network security. This new approach needs to cover a broad range of considerations, including:

- New ways to enforce network security (which workloads are allowed to talk to which other workloads) that do not rely on special meanings of IP addresses or network topology and that work even if the traffic does not traverse the underlying network; the Kubernetes network policy is designed to meet these needs.
- New tools to help manage network policies that support new development processes and the desire for microservices to bring increased organizational agility, such as policy recommendations, policy impact previews, and policy staging.
- New ways to monitor and visualize network traffic, covering both cluster-scoped holistic views (e.g., how to easily view the overall network and the cluster's network security status) and targeted topographic views to drill down across a sequence of microservices to help troubleshoot or diagnose application issues.
- New ways of implementing intrusion detection and threat defense, including policy violation alerting, network anomaly detection, and integrated threat feeds.
- New remediation workflows, so potentially compromised workloads can be quickly and safely isolated during forensic investigation.

- New mechanisms for auditing configuration and policy changes for compliance.
- New mechanisms for auditing configuration and policy changes, and also Kubernetes-aware network flow logs to meet compliance requirements (since traditional network flow logs are IP-based and have little long-term meaning in the context of Kubernetes).

We will review an example of a typical Kubernetes deployment in an enterprise to understand these challenges. Figure 1-2 is a representation of a common deployment model for Kubernetes and microservices in a multicloud environment. A multicloud environment is one where an enterprise deploys Kubernetes in more than one cloud provider (Amazon Web services, Google Cloud, etc.). A hybrid cloud environment is one where an enterprise has a Kubernetes deployment in at least one cloud provider environment and a Kubernetes deployment on-premise in its datacenter. Most enterprises have a dual cloud strategy and will have clusters running in Amazon Web Services (AWS), Microsoft Azure, or Google Cloud; more enterprises also have some legacy applications running in their datacenters. Workloads in the datacenter will likely be behind a security gateway that filters traffic coming in through the perimeter. Microservices running in these Kubernetes deployments are also likely to have one or more dependencies on:

- Other cloud services like AWS RDS or Azure DB
- Third-party API endpoints like Twilio
- SaaS services like Salesforce or Zuora
- Databases or legacy apps running inside the datacenter

Workloads in the datacenter will likely be behind a security gateway that filters traffic coming in through the perimeter.

Observability in Kubernetes is the ability to derive actionable insights about the state of Kubernetes from metrics collected (more on this later). While observability has other applications, like monitoring and troubleshooting, it is important in the context of network security too. Observability concepts applied to flow logs correlated with other Kubernetes metadata (pods labels, policies, namespaces, etc.) are used to monitor (and then secure) communications between pods in a Kubernetes cluster, detect malicious activity by comparing IP addresses with known malicious IP addresses, and use machine learning–based techniques to detect malicious activity. These topics are covered in the next section. As you can see in Figure 1-2, the Kubernetes deployment poses challenges due to silos of data in each cluster and the potential loss of visibility from associating a workload in one cluster to a workload in another cluster or to an external service.

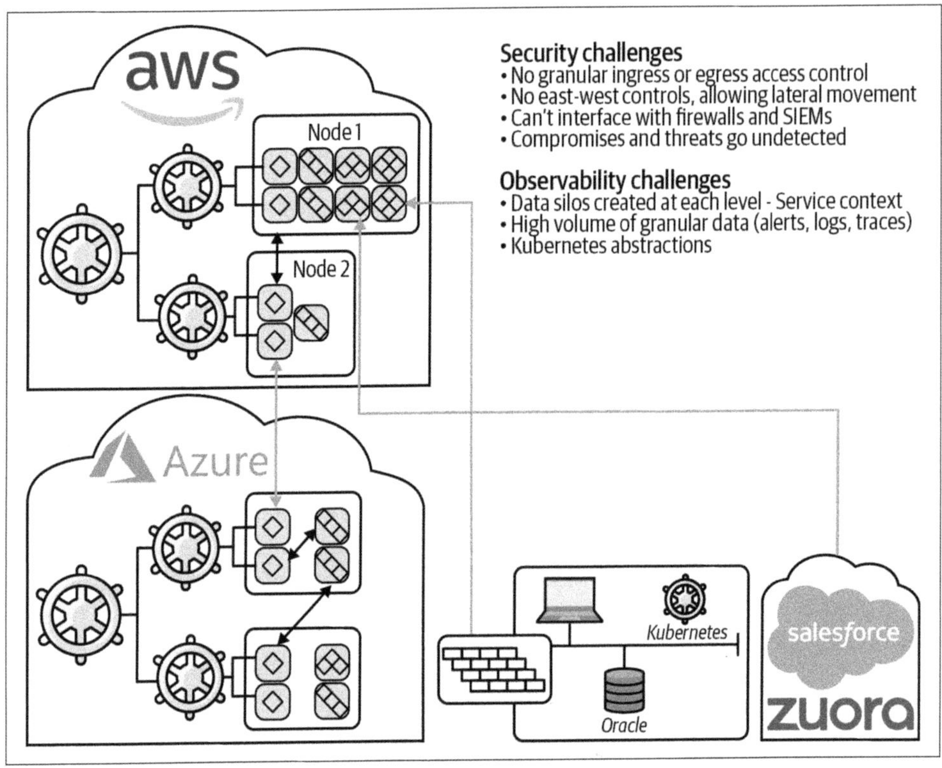

Figure 1-2. Example of a Kubernetes deployment in an enterprise

As shown in Figure 1-2, the footprint of a microservices application typically extends beyond the virtual private cloud (VPC) boundaries, and securing these applications requires a different approach from the traditional perimeter security approach. It is a combination of network security controls, observability, threat defense, and enterprise security controls. We will cover each of these next.

Network security controls

Native security controls available from cloud providers (for example, AWS Security Groups or Azure Network Security Groups) or security gateways (for example, next-generation firewalls) on the perimeter of the VPC or datacenter do not understand the identity of a microservice inside a Kubernetes cluster. For example, you cannot filter traffic to or from a Kubernetes pod or service with your security group rules or firewall policies. Additionally, by the time traffic from a pod hits a cloud provider's network or a third-party firewall, the traffic (depending on the cloud provider's architecture) has a source network address translation (SNAT) applied to it. In other words, the source IP address of traffic from all workloads on the node is set to the node IP, so any kind of allow/deny policies, at best, will have node-level (the node's IP address) granularity.

Kubernetes workloads are highly dynamic and ephemeral. Let's say a developer commits a new check-in for a particular workload. The automated CI/CD workflow will kick in, build a new version of the pod (container), and start deploying this new version of the workload in Kubernetes clusters. Kubernetes orchestrator will do a rolling upgrade and deploy new instances of the workload. All of this happens in an automated fashion, and there is no room for manual or out-of-band workflows to reconfigure the security controls for the newly deployed workload.

You need a new security architecture to secure workloads running in a multi- or hybrid cloud infrastructure. Just like your workload deployment in a Kubernetes cluster, the security of the workload has to be defined as code, in a declarative model. Security controls have to be portable across Kubernetes distributions, clouds, infrastructures, and/or networks. These security controls have to travel with the workloads, so if a new version of the workload is deployed in a VPC for Amazon Elastic Kubernetes Service (EKS), instead of on-premise clusters, you can be assured that the security controls associated with the service will be seamlessly enforced without you having to rework any network topology, out-of-band configuration of security groups, or VPC/perimeter firewalls.

Network security controls are implemented by using a network policy solution that is native to Kubernetes and provides granular access controls. There are several well-known implementations of network policy (such as Calico, Weave Net, Kube-router, Antrea) that you can use. In addition to applying policy at Layer 3/Layer 4 (TCP/IP), we recommend you look at solutions that support application layer policy (such as HTTP/HTTPS). We also recommend picking a solution that is based on the popular proxy Envoy, as it is widely deployed for application-layer policy. Kubernetes supports deploying applications as microservices (small components serving a part of the application functionality) over a network of nodes. The communication between microservices relies on application protocols such as HTTP. Therefore, there is a need for granular application controls that can be implemented by application layer policy. For example, in a three-tier application, the frontend microservice may only be allowed to use HTTP GET-based requests with the backend database microservice (read access) and not allowed to use HTTP POST with the backend database microservice (write access). All these requests can end up using the same TCP connection, so it is essential to add a policy engine that supports application-level controls as described here.

Enterprise security controls

Now that you have the strategy for network access controls and observability defined, you should consider additional security controls that are important and prevalent in enterprises. Encryption of data in transit is a critical requirement for security and compliance. There are several options to consider for encryption using traditional approaches, like TLS-based encryption in your workloads; mutual TLS, which is part of a service mesh platform; or a VPN-based approach like Wireguard (which offers a crypto key–based VPN).

We recommend that you leverage the data collection that is part of your observability strategy to build the reports needed to help with compliance requirements for standards like PCI, HIPAA, GDPR, and SOC 2. You should also consider the ability to ensure continuous compliance, and you can leverage the declarative nature of Kubernetes to help with the design and implementation of continuous compliance. For example, you can respond to a pod failing a compliance check by using the pod's compliance status to trigger necessary action to correct the situation (trigger an image update).

Threat defense

Threat defense in a Kubernetes cluster is the ability to look at malicious activity in the cluster and then defend the cluster from it. Malicious activity allows an adversary to gain unauthorized access and manipulate or steal data from a Kubernetes cluster. The malicious activity can occur in many forms, such as exploiting an insecure configuration or exploiting a vulnerability in the application traffic or the application code.

When you build your threat defense strategy, you must consider both intrusion detection and prevention. The key to intrusion detection is observability; you need to review data collected to scan for known threats. In a Kubernetes deployment, data collection is very challenging due to the large amount of data you need to inspect. We have often heard this question: "Do I need a Kubernetes cluster to collect data to defend a Kubernetes cluster?" The answer is "no." We recommend you align your observability strategy with intrusion detection and leverage smart aggregation to collect and inspect data. For example, you can consider using a tool that aggregates data as groups of "similar" pods talking to each other on a given destination port and protocol, instead of using the traditional method of aggregating by the five-tuple (source IP, source port, destination IP, destination port, protocol). This approach will help significantly reduce data collected without sacrificing effectiveness. Remember, several pods running the same container image and deployed in the same way will generate identical network traffic for a transaction. You may ask, "What if only one instance is infected? How can I detect that?" That's a good question. There are a few ways. You could pick a tool that supports machine learning based on various metrics collected like connections, bytes, and packets to detect anomalous workloads. Another approach is to have a tool that can detect and match known malicious IPs and domains from well-known threat feeds as a part of collection, or log unaggregated network flows for traffic denied by policy. These are simple techniques that will help you build a strategy. Note that threat defense techniques evolve, and you will need a security research team to work with you to help understand your application and build a threat model to implement your threat defense strategy.

Observability

Observability is very useful for monitoring and securing a distributed system like Kubernetes. Kubernetes abstracts a lot of details, and in order to monitor a system like it, you cannot collect and independently baseline and monitor individual metrics (such as a single network flow, a pod create/destroy event, or a CPU spike on one node). What is needed is a way to monitor these metrics in the context of the Kubernetes. For example, a pod associated with a service or a deployment is restarted and running as a different binary as compared to its peers, or a pod activity (network, filesystem, kernel system calls) is different from other pods in the deployment. This becomes even more complex when you consider an application that comprises several services (microservices) that are in turn backed by several pods.

Observability is useful in troubleshooting and monitoring the security of workloads in Kubernetes. As an example, observability in the context of a service in Kubernetes will allow you to do the following:

- Visualize your Kubernetes cluster as a service graph, which shows how pods are associated with services and the communication flows between services
- Overlay application (Layer 7) and network traffic (Layer 3/Layer 4) on the service graph as separate layers that will allow you to easily determine traffic patterns and traffic load for applications and for the underlying network
- View metadata for the node where a pod is deployed (e.g., CPU, memory, or host OS details).
- View metrics related to the operation of a pod, traffic load, application latency (e.g., HTTP duration), network latency (network round-trip time), or pod operation (e.g., RBAC policies, service accounts, or container restarts)
- View DNS activity (DNS response codes, latency, load) for a given service (pods backing the service)
- Trace a user transaction that needs communication across multiple services; this is also known as *distributed tracing*
- View network communication of a given service to external entities
- View Kubernetes activity logs (e.g., audit logs) for pods and resources associated with a given service.

We will cover the details of observability and examples of how it can help security in subsequent chapters. For this discussion, we will cover a brief description of how you can use observability as a part of your security strategy.

Network traffic visibility

As mentioned, a solution that provides network flows aggregated at a service level with context like namespaces, labels, service accounts, or network policies is required to adequately monitor activity and access controls applied to the cluster. For example, there is a significant difference between reporting that IP1 communicated with IP2 over port 8080 and reporting that pods labeled "frontend" communicated with pods labeled "backend" on certain ports or traffic patterns between deployments of pods in a Kubernetes cluster. This reporting will allow you to review communication from external entities and apply IP address–based threat feeds to detect activity from known malicious IP addresses or even traffic from unexpected geographical locations. We will cover details for these concepts in Chapter 11.

DNS activity logs

Domain Name System (DNS) is a system used to translate domain names into IP addresses. In your Kubernetes cluster, it is critical to review DNS activity logs to detect unexpected activity, for example queries to known malicious domains, DNS response codes like NXDOMAIN, and unexpected increases in bytes and packets in DNS queries. We will cover details for these concepts in Chapter 11.

Application traffic visibility

We recommend you review application traffic flows for suspicious activity like unexpected response codes and rare or known malicious HTTP headers (user-agent, query parameters). HTTP is the most common protocol used in Kubernetes deployments, so it is important to work with your security research team to monitor HTTP traffic for malicious traffic. In case you use other application protocols (e.g., Kafka, MySQL), you need to do the same for those as well.

Kubernetes activity logs

In addition to network activity logs, you must also monitor Kubernetes activity logs to detect malicious activity. For example, review access-denied logs for resources access and service account creation/modification. Review namespace creation/modification logs for unexpected activity. And review the Kubernetes audit logs which record requests to the Kubernetes API.

Machine learning/anomaly detection

Machine learning is a technique where a system is able to derive patterns from data over a period of time. The output is a machine learning model, which can then be used to make predictions and detect deviations in real data based on the prediction. We recommend you consider applying machine learning–based anomaly detection to various metrics to detect strange activity. A simple and effective way is to apply a machine learning technique known as *baselining* to individual metrics. This way you do not need to worry about applying rules and thresholds for each metric; the system does that for you and reports deviations as anomalies. Applying machine learning techniques to network traffic is a relatively new area and is gaining traction with security teams. We will cover this topic in detail in Chapter 6.

There are many solutions that you can choose for your observability strategy for Kubernetes (Datadog, Calico Enterprise, cloud provider–based solutions from Google, AWS, Azure).

Security Frameworks

Finally, we want to make you aware of security frameworks that provide the industry a common methodology and terminology for security best practices. Security frameworks are a great way to understand attack techniques and best practices to defend and mitigate attacks. You should use them to build and validate your security strategy. Please note these frameworks may not be specific to Kubernetes, but they provide insights into techniques used by adversaries in attacks, and security researchers will need to review and see if they are relevant to Kubernetes. We will review two well-known frameworks—MITRE and Threat Matrix for Kubernetes.

MITRE

MITRE is a knowledge base of adversary tactics and techniques based on real-world observations of cyberattacks. The MITRE ATT&CK® Matrix for Enterprise (*https://oreil.ly/fxBKB*) is useful because it provides the tactics and techniques categorized for each stage of the cybersecurity kill chain. The kill chain is a description of the stages in a cyberattack and is useful for building an effective defense against an attack. MITRE also provides an attack matrix tailored for cloud environments like AWS, Google Cloud, and Microsoft Azure.

Figure 1-3 describes the MITRE ATT&CK® Matrix for AWS. (*https://oreil.ly/Mvyzz*) We recommend that you review each of the stages described in the attack matrix as you build your threat model for securing your Kubernetes cluster.

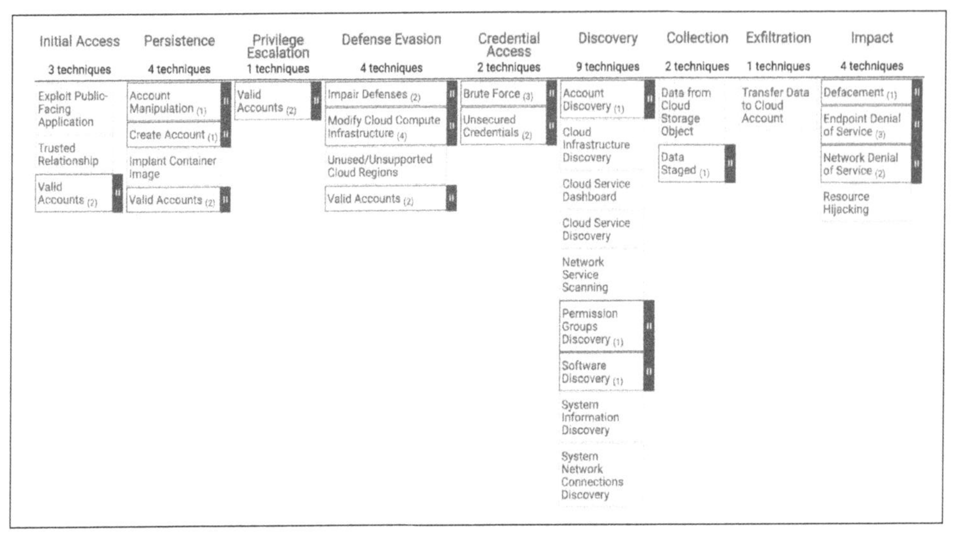

Figure 1-3. Attack matrix for cloud environments in AWS

Threat matrix for Kubernetes

The other framework is a threat matrix (*https://oreil.ly/GQery*) that is a Kubernetes-specific application of the generic MITRE attack matrix. It was published by the Microsoft team based on security research and real-world attacks. This is another excellent resource to use to build and validate your security strategy.

Figure 1-4 provides the stages that are relevant to your Kubernetes cluster. They map to the various stages we discussed in this chapter. For example, you should consider the compromised images in the registry in the initial access stage, the access cloud resources in the privilege escalation stage, and the cluster internal network in the lateral movement stage for build, deploy, and runtime security, respectively.

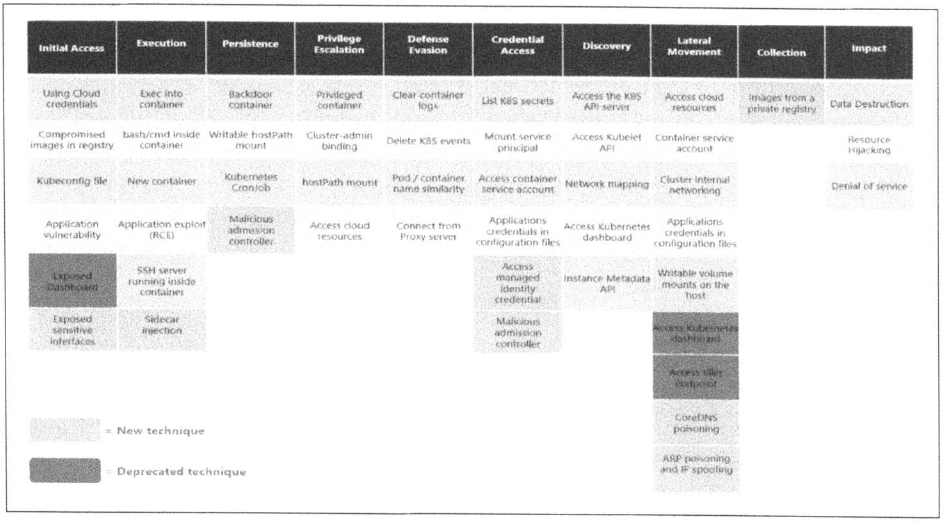

Initial Access	Execution	Persistence	Privilege Escalation	Defense Evasion	Credential Access	Discovery	Lateral Movement	Collection	Impact
Using Cloud credentials	Exec into container	Backdoor container	Privileged container	Clear container logs	List K8S secrets	Access the K8S API server	Access cloud resources	Images from a private registry	Data Destruction
Compromised images in registry	bash/cmd inside container	Writable hostPath mount	Cluster-admin binding	Delete K8S events	Mount service principal	Access Kubelet API	Container service account		Resource Hijacking
Kubeconfig file	New container	Kubernetes CronJob	hostPath mount	Pod / container name similarity	Access container service account	Network mapping	Cluster internal networking		Denial of service
Application vulnerability	Application exploit (RCE)	Malicious admission controller	Access cloud resources	Connect from Proxy server	Applications credentials in configuration files	Access Kubernetes dashboard	Applications credentials in configuration files		
Exposed Dashboard	SSH server running inside container				Access managed identity credential	Instance Metadata API	Writable volume mounts on the host		
Exposed sensitive interfaces	Sidecar injection				Malicious admission controller		Access Kubernetes dashboard		
							Access tiller endpoint		
							CoreDNS poisoning		
							ARP poisoning and IP spoofing		

= New technique

= Deprecated technique

Figure 1-4. Threat matrix for Kubernetes

Security and Observability

In a dynamic environment like Kubernetes, a secure deployment of your applications can be achieved by thinking about security and observability together. As an example, you need to "observe" your cluster to find the optimal way to implement controls to secure the cluster. Kubernetes as an orchestration engine has strong adoption due to the fact that it is declarative in nature, allowing users to specify higher-level outcomes. Kubernetes also has built-in capabilities to ensure that your cluster operates as per the specifications. It does this by monitoring the various attributes and taking action (e.g., a pod restart) if the attribute deviates from the specified value for a period of time. These aspects of Kubernetes make it difficult to implement the visibility and controls needed to secure a cluster. The controls you implement need to be aligned with Kubernetes operations. Therefore, before you think of adding any controls to Kubernetes, it is important to understand the context—for example, you cannot isolate a pod by applying a policy that does not allow it to communicate with anything else. Kubernetes will detect that the pod is not able to communicate with the other elements (e.g., API server), determine that the pod is not operating as specified, and restart and spin up the pod somewhere else in the cluster.

What you have to do is first understand how the pod operates and understand what its expected operation is and then apply controls or detect unexpected events. After that, you determine if the unexpected event is an operations issue or a security issue and then apply the required remediation. In order to do this, observability and security go hand in hand: You observe to understand what is expected and apply controls to ensure expected operation, then observe to detect unexpected events and analyze them, and then add necessary controls to remediate any issue due to the event. Therefore, you need a holistic approach for security and observability when you think about securing your clusters.

Conclusion

By now you should have a high-level overview of what Kubernetes security and observability entails. These are the foundational concepts that underpin this entire book. In short:

- Security for Kubernetes is very different from traditional security and requires a holistic security and observability approach at all the stages of workload deployment—build, deploy, and runtime.
- Kubernetes is declarative and abstracts the details of workload operations, which means workloads can be running anywhere over a network of nodes. Also, workloads can be ephemeral, where they are destroyed and re-created on a different node. Securing such a declarative distributed system requires that you think about security at all stages.
- We hope you understand the importance of collaboration between the application, platform, and security teams when designing and implementing a holistic security approach.
- MITRE and the Threat Matrix for Kubernetes are two security frameworks that are widely adopted by security teams.

It's important that you take in all of this together, because a successful security and observability strategy is a holistic one. In the next chapter, we will cover infrastructure security.

CHAPTER 2

Infrastructure Security

Many Kubernetes configurations are insecure by default. In this chapter we will explore how to secure Kubernetes at the infrastructure level. It can be made more secure through the combination of host hardening to make the servers or VMs Kubernetes is hosted on more secure, cluster hardening to secure the Kubernetes control plane components, and network security to integrate the cluster with the surrounding infrastructure. Please note that the concepts discussed in this chapter apply to self-hosted Kubernetes clusters as well as managed Kubernetes clusters.

Host hardening
: This covers the choice of operating system, avoiding running nonessential processes on the hosts, and host-based firewalling.

Cluster hardening
: This covers a range of configuration and policy settings needed to harden the control plane, including configuring TLS certificates, locking down the Kubernetes datastore, encrypting secrets at rest, credential rotation, and user authentication and access control.

Network security
: This covers securely integrating the cluster with the surrounding infrastructure, and in particular which network interactions between the cluster and the surrounding infrastructure are allowed, for control plane, host, and workload traffic.

Let's look at the details for each of these aspects and explore what is needed to build a secure infrastructure for your Kubernetes cluster.

Host Hardening

A secure host is an important building block for a secure Kubernetes cluster. When you think of a host, it is in the context of workloads that make up your Kubernetes cluster. We will now explore techniques to ensure a strong security posture for the host.

Choice of Operating System

Many enterprises standardize on a single operating system across all of their infrastructure, which means the choice may have already been made for you. However, if there is flexibility to choose an operating system, then it is worth considering a modern immutable Linux distribution specifically designed for containers. These distributions are advantageous for the following reasons:

- They often have newer kernels, which include the latest vulnerability fixes as well as up-to-date implementations of newer technologies such as eBPF, which can be leveraged by Kubernetes networking and security monitoring tools.
- They are designed to be immutable, which brings additional benefits for security. Immutability in this context means that the root filesystem is locked and cannot be changed by applications. Applications can only be installed using containers. This isolates applications from the root filesystem and significantly reduces the ability for malicious applications to compromise the host.
- They often include the ability to self-update to newer versions, with the upstream versions being geared up for rapid releases to address security vulnerabilities.

Two popular examples of modern immutable Linux distributions designed for containers are Flatcar Container Linux (which was originally based on CoreOS Container Linux) and Bottlerocket (originally created and maintained by Amazon).

Whichever operating system you choose, it is good practice to monitor upstream security announcements so you know when new security vulnerabilities are identified and disclosed and to make sure you have processes in place to update your cluster to a newer version to address critical security vulnerabilities. Based on your assessment of these vulnerabilities, you will want to make a decision on whether to upgrade your cluster to a new version of the operating system. When you consider the choice of the operating system, you must also take into account shared libraries from the host operating system and understand their impact on containers that will be deployed on the host.

Another security best practice is to ensure that application developers do not depend on a specific version of the operating system or kernel, as this will not allow you to update the host operating system as needed.

Nonessential Processes

Each running host process is a potential attack vector for hackers. From a security perspective, it is best to remove any nonessential processes that may be running by default. If a process isn't needed for the successful running of Kubernetes, management of your host, or security of your host, then it is best not to run the process. How you disable the process will depend on your particular setup (e.g., systemd configuration change or removing the initialization script from /etc/init.d/).

If you are using an immutable Linux distribution optimized for containers, then nonessential processes will have already been eliminated and you can only run additional processes/applications as containers.

Host-Based Firewalling

To further lock down the servers or VMs Kubernetes is hosted on, the host itself can be configured with local firewall rules to restrict which IP address ranges and ports are allowed to interact with the host.

Depending on your operating system, this can be done with traditional Linux admin tools such as iptables rules or firewalld configuration. It is important to make sure any such rules are compatible with both the Kubernetes control plane and whichever Kubernetes network plug-in you plan to use so they do not block the Kubernetes control plane, pod networking, or the pod network control plane. Getting these rules right, and keeping them up to date over time, can be a time-consuming process. In addition, if using an immutable Linux distribution, you may not easily be able to directly use these tools.

Fortunately, some Kubernetes network plug-ins can help solve this problem for you. For example, several Kubernetes network plug-ins, like Weave Net, Kube-router, and Calico, include the ability to apply network policies. You should review these plug-ins and pick one that also supports applying network policies to the hosts themselves (rather than just to Kubernetes pods). This makes securing the hosts in the cluster significantly simpler and is largely operating system independent, including working with immutable Linux distributions.

Always Research the Latest Best Practices

As new vulnerabilities or attack vectors are identified by the security research community, security best practices evolve over time. Many of these are well-documented online and are available for free.

For example, the Center for Internet Security maintains free PDF guides with comprehensive configuration guidance to secure many of the most common operating systems. Known as CIS Benchmarks, they are an excellent resource for making sure you are covering the many important actions required to secure your host operating system. You can find an up-to-date list of CIS Benchmarks on their website (*https://oreil.ly/dpUnC*). Please note there are Kubernetes-specific benchmarks, and we will discuss them later in this chapter.

Cluster Hardening

Kubernetes is insecure by default. So in addition to hardening the hosts that make up a cluster, it is important to harden the cluster itself. This can be done through a combination of Kubernetes component and configuration management, authentication and role-based access control (RBAC), and keeping the cluster updated with the latest versions of Kubernetes to ensure the cluster has the latest vulnerability fixes.

Secure the Kubernetes Datastore

Kubernetes uses etcd as its main datastore. This is where all cluster configuration and desired state is stored. Having access to the etcd datastore is essentially equivalent to having root login on all your nodes. Almost any other security measures you have put in place within the cluster become moot if a malicious actor gains access to the etcd datastore. They will have complete control over your cluster at that point, including the ability to run arbitrary containers with elevated privileges on any node.

The main way to secure etcd is to use the security features provided by etcd itself. These are based around x509 Public Key Infrastructure (PKI), using a combination of keys and certificates. They ensure that all data in transit is encrypted with TLS and all access is restricted with strong credentials. It is best to configure etcd with one set of credentials (key pairs and certificates) for peer communications between the different etcd instances, and another set of credentials for client communications from the Kubernetes API. As part of this configuration, etcd must also be configured with the details of certificate authority (CA) used to generate the client credentials.

Once etcd is configured correctly, only clients with valid certificates can access it. You must then configure the Kubernetes API server with the client certificate, key, and certificate authority so it can access etcd.

You can also use network-level firewall rules to restrict etcd access so it can only be accessed from Kubernetes control nodes (hosting the Kubernetes API server). Depending on your environment you can use a traditional firewall, virtual cloud firewall, or rules on the etcd hosts themselves (for example, a networking policy implementation that supports host endpoint protection) to block traffic. This is best done in addition to using etcd's own security features as part of an in-depth defense strategy, since limiting access with firewall rules does not address the security need for Kubernetes' sensitive data to be encrypted in transit.

In addition to securing etcd access for Kubernetes, it is recommended to not use the Kubernetes etcd datastore for anything other than Kubernetes. In other words, do not store non-Kubernetes data within the datastore and do not give other components access to the etcd cluster. If you are running applications or infrastructure (within the cluster or external to the cluster) that uses etcd as a datastore, the best practice is to set up a separate etcd cluster for that. The arguable exception would be if the application or infrastructure was sufficiently privileged that a compromise to its datastore would also result in a complete compromise of Kubernetes. It is also very important to maintain backups of etcd and secure the backups so that it is possible to recover from failures like a failed upgrade or a security incident.

Secure the Kubernetes API Server

One layer up from the etcd datastore, the next set of crown jewels to be secured is the Kubernetes API server. As with etcd, this can be done using x509 PKI and TLS. The details of how to bootstrap a cluster in this way vary depending on the Kubernetes installation method you are using, but most methods include steps that create the required keys and certificates and distribute them to the other Kubernetes cluster components. It's worth noting that some installation methods may enable insecure local ports for some components, so it is important to familiarize yourself with the settings of each component to identify potential unsecured traffic so you can take appropriate action to secure them.

Encrypt Kubernetes Secrets at Rest

Kubernetes can be configured to encrypt sensitive data it stores in etcd, such as Kubernetes secrets. This keeps the secrets safe from any attacker that may gain access to etcd or to an offline copy of etcd such as offline backup.

By default, Kubernetes does not encrypt secrets at rest, and when encryption is enabled, it only encrypts when a secret is written to etcd. Therefore, when enabling encryption at rest, it is important to rewrite all secrets (through standard kubectl apply or update commands) to trigger their encryption within etcd.

Kubernetes supports a variety of encryption providers. It is important to pick the recommended encryption based on encryption best practices. The mainline recommended choice is AES-CBC with PKCS #7–based encryption. This provides very strong encryption using 32-byte keys and is relatively fast. There are two different providers that support this encryption:

- The local provider that runs entirely with Kubernetes and uses locally configured keys
- The KMS provider that uses an external key management service (KMS) to manage the keys

The local provider stores its keys on the API server's local disk. This therefore has the limitation that if the API server host is compromised, then all of your secrets become compromised. Depending on your security posture, this may be acceptable.

The KMS provider uses a technique called *envelope encryption*. With envelope encryption, each secret is encrypted with a dynamically generated data encryption key (DEK). The DEK is then encrypted using a key encryption key (KEK) provided by the KMS, and the encrypted DEK is stored alongside the encrypted secret in etcd. The KEK is always hosted by the KMS as the central root of trust and is never stored within the cluster. Most large public cloud providers offer a cloud-based KMS service that can be used as the KMS provider in Kubernetes. For on-prem clusters there are third-party solutions, such as HashiCorp's Vault, which can act as the KMS provider for the cluster. Because the detailed implementations vary, it is important to evaluate the mechanism through which the KMS authenticates the API server and whether a compromise to the API server host could in turn compromise your secrets and therefore offer only limited benefits compared with a local encryption provider.

If exceptionally high volumes of encrypted storage read/writes are anticipated, then using the secretbox encryption provider could potentially be faster. However, secretbox is a newer standard and at the time of writing has had less review than other encryption algorithms. It therefore may not be considered acceptable in environments that require high levels of review. In addition, secretbox is not yet supported by the KMS provider and must use a local provider, which stores the keys on the API server.

Encrypting Kubernetes secrets is the most common must-have encryption at-rest requirement, but note that you can also configure Kubernetes to encrypt storage of other Kubernetes resources if desired.

It's also worth noting there are third-party secret management solutions that can be used if you have requirements beyond the capabilities of Kubernetes secrets. One such solution, already mentioned as a potential KMS provider for envelope encryption in Kubernetes, is HashiCorp's Vault. In addition to providing secure secrets management for Kubernetes, Vault can be used beyond the scope of Kubernetes to manage secrets more broadly across the enterprise if desired. Vault was also a very popular choice for plugging the major gap in earlier Kubernetes versions, which did not support the encryption of secrets at rest.

Rotate Credentials Frequently

Rotating credentials frequently makes it harder for attackers to make use of any compromised credential they may obtain. It is therefore a best practice to set short lifetimes on any TLS certificates or other credentials and automate their rotation. Most authentication providers can control how long issued certificates or service tokens are valid for, and it is best to use short lifetimes whenever possible, for example rotating keys every day or more frequently if they are particularly sensitive. This needs to include any service tokens used in external integrations or as part of the cluster bootstrap process.

Fully automating the rotation of credentials may require custom DevOps development work, but normally represents a good investment compared with attempting to manually rotate credentials on an ongoing basis.

When rotating keys for Kubernetes secrets stored at rest (as discussed in the previous section), local providers support multiple keys. The first key is always used to encrypt any new secret writes. For decryption, the keys are tried in order until the secret is successfully decrypted. As keys are encrypted only on writes, it is important to rewrite all secrets (through standard kubectl apply or update commands) to trigger their encryption with the latest keys. If the rotation of secrets is fully automated, then the write will happen as part of this process without requiring a separate step.

When using a KMS provider (rather than a local provider), the KEK can be rotated without requiring reencryption of all the secrets, which can reduce the performance impact of reencrypting all secrets if you have a large number of sizable secrets.

Authentication and RBAC

In the previous sections we primarily focused on securing programmatic/code access within the cluster. Equally important is to follow best practices for securing user interactions with the cluster. This includes creating separate user accounts for each user and using Kubernetes RBAC to grant users the minimal access they need to perform their role, following the principle of least privilege access. Usually it is better to do this using groups and roles, rather than assigning RBAC permissions to individual users. This makes it easier to manage privileges over time, both in terms of adjusting privileges for different groups of users when requirements change and for reducing the effort required to periodically review/audit the user privileges across the cluster to verify they are correct and up to date.

Kubernetes has limited built-in authentication capabilities for users, but can be integrated with most external enterprise authentication providers, such as public cloud provider IAM systems or on-prem authentication services, either directly or through third-party projects such as Dex (originally created by CoreOS). It is generally recommended to integrate with one of these external authentication providers rather than using Kubernetes basic auth or service account tokens, since external authentication providers typically have more user-friendly support for rotation of credentials, including the ability to specify password strength and rotation frequency timeframes.

Restricting Cloud Metadata API Access

Most public clouds provide a metadata API that is accessible locally from each host/VM instance. The APIs provide access to the instance's cloud credentials, IAM permissions, and other potentially sensitive information about the instance. By default, these APIs are accessible by the Kubernetes pods running on an instance. Any compromised pod can use these credentials to elevate its intended privilege level within the cluster or to other cloud provider–hosted services the instance may have privileges to access.

To address this security issue, the best practice is to:

- Provide any required pod IAM credentials following the cloud provider's recommended mechanisms. For example, Amazon EKS allows you to assign a unique IAM role to a service account, Microsoft Azure's AKS allows you to assign a managed identity to a pod, and Google Cloud's GKE allows you to assign IAM permissions via Workload Identity.
- Limit the cloud privileges of each instance to the minimum required to reduce the impact of any compromised access to the metadata API from the instance.
- Use network policies to block pod access to the metadata API. This can be done with per-namespace Kubernetes network policies, or preferably with extensions to Kubernetes network policies such as those offered by Calico, which enable a single network policy to apply across the whole of the cluster (without the need to create a new Kubernetes network policy each time a new namespace is added to the cluster). This topic is covered in more depth in the Default Deny and Default App Policy section of Chapter 7.

Enable Auditing

Kubernetes auditing provides a log of all actions within the cluster with configurable scope and levels of detail. Enabling Kubernetes audit logging and archiving audit logs on a secure service is recommended as an important source of forensic details in the event of needing to analyze a security breach.

The forensic review of the audit log can help answer questions such as:

- What happened, when, and where in the cluster?
- Who or what initiated it and from where?

In addition, Kubernetes audit logs can be actively monitored to alert on suspicious activity using your own custom tooling or third-party solutions. There are many enterprise products that you can use to monitor Kubernetes audit logs and generate alerts based on configurable match criteria.

The details of what events are captured in Kubernetes audit logs are controlled using policy. The policy determines which events are recorded, for which resources, and with what level of detail.

Each action being performed can generate a series of events, defined as stages:

RequestReceived
: Generated as soon as the audit handler receives the request

ResponseStarted
: Generated when the response headers are sent, but before the response body is sent (generated only for long-running requests such as watches)

ResponseComplete
: Generated when the response body has been completed

Panic
: Generated when a panic occurs

The level of detail recorded for each event can be one of the following:

None
: Does not log the event at all

Metadata
: Logs the request metadata (user, timestamp, resource, verb, etc.) but not the request details or response body

Request
: Logs event metadata and the request body but not the response body

RequestResponse
: Logs the full details of the event, including the metadata and request and response bodies

Kubernetes' audit policy is very flexible and well documented in the main Kubernetes documentation. Included here are just a couple of simple examples to illustrate.

To log all requests at the metadata level:

```
apiVersion: audit.k8s.io/v1
kind: Policy
rules:
- level: Metadata
```

To omit the RequestReceived stage and to log pod changes at the RequestResponse level and configmap and secrets changes at the metadata level:

```
apiVersion: audit.k8s.io/v1 # This is required.
kind: Policy
omitStages:
  - "RequestReceived"
rules:
  - level: RequestResponse
    resources:
    - group: ""
      resources: ["pods"]
  - level: Metadata
    resources:
    - group: "" # core API group
      resources: ["secrets", "configmaps"]
```

This second example illustrates the important consideration of sensitive data in audit logs. Depending on the level of security around access to the audit logs, it may be essential to ensure the audit policy does not log details of secrets or other sensitive data. Just like the practice of encrypting secrets at rest, it is generally a good practice to always exclude sensitive details from your audit log.

Restrict Access to Alpha or Beta Features

Each Kubernetes release includes alpha and beta features. Whether these are enabled can be controlled by specifying feature gate flags for the individual Kubernetes components. As these features are in development, they can have limitations or bugs that result in security vulnerabilities. So it is a good practice to make sure that all alpha and beta features you do not intend to use are disabled.

Alpha features are normally (but not always) disabled by default. They might be buggy, and the support for the feature could radically change without backward compatibility, or be dropped in future releases. They are generally recommended for testing clusters only, not production clusters.

Beta features are normally enabled by default. They can still change in non-backward-compatible ways between releases, but they are usually considered reasonably well tested and safe to enable. But as with any new feature, they are inherently more likely to have vulnerabilities because they have been used less and had less review.

Always assess the value an alpha or beta feature may provide against the security risk it represents, as well as the potential ops risk of non-backward-compatible changes of the features between releases.

Upgrade Kubernetes Frequently

It is inevitable that new vulnerabilities will be discovered over time in any large software project. The Kubernetes developer community has a good track record of responding to newly discovered vulnerabilities in a timely manner. Severe vulnerabilities are normally fixed under embargo, meaning that the knowledge of the vulnerability is not made public until the developers have had time to produce a fix. Over time, the number of publicly known vulnerabilities for older Kubernetes versions grows, which can put older clusters at greater security risk.

To reduce the risk of your clusters being compromised, it is important to regularly upgrade your clusters, and to have in place the ability to urgently upgrade if a severe vulnerability is discovered.

All Kubernetes security updates and vulnerabilities are reported (once any embargo ends) via the public and free-to-join *kubernetes-announce* email group. Joining this group is highly recommended for anyone wanting to keep track of known vulnerabilities so they can minimize their security exposure.

Use a Managed Kubernetes Service

One way to reduce the effort required to act on all of the advice in this chapter is to use one of the major public cloud managed Kubernetes services, such as EKS, AKS, GKE, or IKS. Using one of these services moves security from being 100% your own responsibility to being a shared responsibility model. Shared responsibility means there are many elements of security that the service includes by default, or can be easily configured to support, but that there are elements you still have to take responsibility for yourself in order to make the whole truly secure.

The details vary depending on which public cloud service you are using, but there's no doubt that all of them do significant heavy lifting, which reduces the effort required to secure the cluster compared to if you are installing and managing the cluster yourself. In addition, there are plenty of resources available from the public cloud providers and from third parties that detail what you need to do as part of the shared responsibility model to fully secure your cluster. For example, one of these resources is CIS Benchmarks, as discussed next.

CIS Benchmarks

As discussed earlier in this chapter, CIS maintains free PDF guides with comprehensive configuration guidance to secure many of the most common operating systems. These guides, known as CIS Benchmarks, can be an invaluable resource to help you with host hardening.

In addition to helping with host hardening, there are also CIS Benchmarks for Kubernetes itself, including configuration guidance for many of the popular managed Kubernetes services, which can help you implement much of the guidance in this chapter. For example, the GKE CIS Benchmark includes guidance on ensuring the cluster is configured with autoupgrade of nodes, and for using managing Kubernetes authentication and RBAC with Google Groups. These guides are highly recommended resources to keep up to date with the latest practical advice on the steps required to secure Kubernetes clusters.

In addition to the guides themselves, there are third-party tools available that can assess the status of a running cluster against many of these benchmarks. One popular tool is kube-bench, an open source project originally created by the team at Aqua Security. Or if you prefer a more packaged solution, then many enterprise products have CIS Benchmark and other security compliance tooling and alerting built into their cluster management dashboards. Having these kinds of tools in place, ideally running automatically at regular intervals, can be valuable for verifying the security posture of a cluster and ensuring that careful security measures that might have been put in place at cluster creation time are not accidentally lost or compromised as the cluster is managed and updated over time.

Network Security

When securely integrating the cluster with the surrounding infrastructure outside the scope of the cluster, network security is the primary consideration. There are two aspects to consider: how to protect the cluster from attack from outside the cluster and how to protect the infrastructure outside of the cluster from any compromised element inside the cluster. This applies at both the cluster workload level (i.e., Kubernetes pods) and the cluster infrastructure level (i.e., the Kubernetes control plane and the hosts on which the Kubernetes cluster is running).

The first thing to consider is whether or not the cluster needs to be accessible from the public internet, either directly (e.g., one or more nodes have public IP addresses) or indirectly (e.g., via a load balancer or similar that is reachable from the internet). If the cluster is accessible from the internet, then the number of attacks or probes from hackers is massively increased. So if there is not a strong requirement for the cluster to be accessible from the internet, then it is highly recommended to not allow any access at a routability level (i.e., ensuring there is no way of packets getting from the internet to the cluster). In an enterprise on-prem environment, this may equate to the choice of IP address ranges to use for the cluster and their routability within the enterprise network. If using a public cloud managed Kubernetes service, you may find these settings can be set only at cluster creation. For example, in GKE, whether the Kubernetes control plane is accessible from the internet can be set at cluster creation.

Network policy within the cluster is the next line of defense. Network policy can be used to restrict both workload and host communications to/from the cluster and the infrastructure outside of the cluster. It has the strong advantage of being workload aware (i.e., the ability to limit communication of groups of pods that make up an individual microservices) and being platform agnostic (i.e., the same techniques and policy language can be used in any environment, whether on-prem within the enterprise or in public cloud). Network policy is discussed in depth later in a dedicated chapter.

Finally, it is highly recommended to use perimeter firewalls, or their cloud equivalents such as security groups, to restrict traffic to/from the cluster. In most cases these are not Kubernetes workload aware, so they don't understand individual pods and are therefore usually limited in granularity to treating the whole of the cluster as a single entity. Even with this limitation they add value as part of a defense strategy, though on their own they are unlikely to be sufficient for any security-conscious enterprise.

If stronger perimeter defense is desired, there are strategies and third-party tools that can make perimeter firewalls or their cloud equivalents more effective:

- One approach is to designate a small number of specific nodes in the cluster as having a particular level of access to the rest of the network, which is not granted to the rest of the nodes in the cluster. Kubernetes taints can then be used to ensure that only workloads that need that special level of access are scheduled to those nodes. This way perimeter firewall rules can be set based on the IP addresses of the specific nodes to allow desired access, and all other nodes in the cluster are denied access outside the cluster.
- In an on-prem environment, some Kubernetes network plug-ins allow you to use routable pod IP addresses (nonoverlay networks) and control the IP address ranges that the group of pods backing a particular microservice use. This allows perimeter firewalls to act on IP address ranges in a similar way as they do with traditional non-Kubernetes workloads. For example, you need to pick a network plug-in that supports nonoverlay networks on-prem that are routable across the broader enterprise networks and that has flexible IP address management capabilities to facilitate such an approach.
- A variation of the previous is useful in any environment where it is not practical to make pod IP addresses routable outside of the cluster (e.g., when using an overlay network). In this scenario, the network traffic from pods appears to come from the IP address of the node as it uses source network address translation (SNAT). In order to address this, you can use a Kubernetes network plug-in that supports fine-grained control of egress NAT gateways. The egress NAT gateway feature supported by some Kubernetes network plug-ins allows this behavior to be changed so that the egress traffic for a set of pods is routed via specific gateways within the cluster that perform the SNAT, so the traffic appears to be

coming from the gateway, rather than from the node hosting the pod. Depending on the network plug-in being used, the gateways can be allocated to specific IP address ranges or to specific nodes, which in turn allows perimeter firewall rules to act more selectively than treating the whole of the cluster as a single entity. There are a few options that support this functionality: Red Hat's OpenShift SDN, Nirmata, and Calico all support egress gateways.

- Finally, some firewalls support some plug-ins or third-party tools that allow the firewall to be more aware of Kubernetes workloads, for example, by automatically populating IP address lists within the firewall with pod IP addresses (or node IP addresses of the nodes hosting particular pods). Or in a cloud environment, there are automatic programming rules that allow security groups to selectively act on traffic to/from Kubernetes pods, rather than operating only at the node level. This integration is very important for your cluster to help complement the security provided by firewalls. There are several tools in the market that allow this type of integration. It is important to choose a tool that supports these integrations with major firewall vendors and is native to Kubernetes.

In the previous section we discussed the importance of network security and how you can use network security to secure access to your Kubernetes cluster from traffic originating outside the cluster and also how to control access for traffic originating from within the cluster destined to hosts outside the cluster.

Conclusion

In this chapter we discussed the following key concepts, which you should use to ensure you have a secure infrastructure for your Kubernetes cluster.

- You need to ensure that the host is running an operating system that is secure and free from critical vulnerabilities.
- You need to deploy access controls on the host to control access to the host operating system and deploy controls for network traffic to and from the host.
- You need to ensure a secure configuration for your Kubernetes cluster; securing the datastore and API server are key to ensuring you have a secure cluster configuration.
- Finally, you need to deploy network security to control network traffic that originates from pods in the cluster and is destined to pods in the cluster.

CHAPTER 3

Workload Deployment Controls

With contributions from Manoj Ahuje,
Senior Threat Intelligence Research Engineer at Tigera

Once you decide on a strategy for infrastructure security, next in line is workload deployment controls. In this chapter we will look at image building and scanning strategy, CI/CD (integrating image scanning into builds), and Kubernetes role-based access control (RBAC), which is a widely used authorization system that allows you to define access control based on user roles, and secrets management for your applications.

Image Building and Scanning

In this section we will explore best practices for image building and scanning. These include choosing a base image to reduce attack surface and using scratch images and image hardening best practices to deter adversaries. Image scanning dives into the nuances of choosing an image scanning solution, privacy concerns, and an overview of container threat analysis solutions.

Choice of a Base Image

As discussed in the previous chapter, you can choose modern Linux distributions like Bottlerocket as base images for containers. The minimal version of traditional Linux distributions, like Ubuntu, Red Hat, and Alpine, are available too.

Though it's a good starting point to begin with a minimal image, the minimal image approach doesn't stop vulnerabilities being discovered in OS packages that are present in the OS. In this case distroless or scratch images turn out to be a better option. These types of images only contain the application and its specific runtime dependencies.

Here are the benefits of distroless or scratch images:

- This strategy reduces size, attack surface, and vulnerabilities significantly, which results in better security posture.
- Distroless images are production ready. Kubernetes itself uses distroless images for various components like kublet, scheduler, etc.
- In case you are going for a scratch base image for your application, a multistage Dockerfile can be used to build a scratch image. The first stage involves building your application. The second stage involves moving runtime dependencies and applications to scratch.

The most popular example of a distroless image project is distroless from Google (*https://oreil.ly/UIzu0*), which provides images for various runtimes like Java, Python, or C++.

A scratch image starts with the Dockerfile instruction FROM:scratch, which signifies an empty filesystem. The following instruction in Dockerfile creates the first filesystem layer of the container image. Here, the first filesystem layer needs to be compiled with the application and dependencies. Since it's nonproductive and nonintuitive to build applications outside of the container, Docker introduced multistage builds (*https://oreil.ly/K161o*). With a multistage build, multiple FROM instructions are allowed in a Dockerfile. Each FROM instruction creates a separate stage, and the filesystem artifact from a previous stage can be copied in a later stage of the build. This mechanism enabled developers to build and compile applications in an earlier stage (builder image) with all dependencies available, and eventually only copy the filesystem artifacts required to run the production application in a later stage. The last stage of the build can be a scratch image where only application binaries and dependencies are required to be present on the resulting image.

Following is an example of a scratch-based image for a bash script, where the goal is to run a script within a container. Here you can create a two-stage Dockerfile where the second stage is a scratch image containing only the dependencies for the script.

You can use this template to containerize even complex applications built with Node.js, Python, and Go. Go additionally provides an option to compile all the runtime libraries into the binary. In the following example, you can use Alpine as the base image to construct a scratch image for a container that runs a script:

```
# use alpine 3 as base image
FROM alpine:3 as builder

# upgrade all alpine packages
RUN apk update && apk upgrade

# add your script into container fs
ADD your_init_script.sh your_init_script.sh
RUN chmod u+x your_init_script.sh

# stage 2
FROM scratch

# shell
COPY --from=builder /bin/sh /bin/sh

# dependent linux shared libraries
COPY --from=builder /lib/ld-musl-x86_64.so.1 /lib/ld-musl-x86_64.so.1

ENV PATH=/usr/local/bin:/usr/local/sbin:/usr/local/bin:/usr/sbin:/usr/bin:
/sbin:/bin

ENTRYPOINT ["./your_init_script.sh"]
In this example, only two files are copied into scratch images (second stage).
Hence, the alpine base image containing more than five thousand files is minimized
to two files, reducing the attack surface significantly.
```

Now that we have reviewed how to choose a base image for your container, let's explore container image hardening.

Container Image Hardening

Container image hardening is the process of building images to reduce security weaknesses and attack surface. At the same time, it is used to add defensive layers to run applications securely within the container.

If you use a nonhardened container image, it can be prone to abuse, information disclosure, or easier privilege escalation to the container host. You should leverage the following tools and best practices to build hardened container images for your applications:

- Only use base images from trusted sources, such as official Ubuntu and Red Hat release channels, and double-check the image hash with released information, as it's relatively easy to embed malicious code like cryptominers into an image and make that image available on a repository like Docker Hub.
- Minimize your base images to only contain runtime dependencies for your application.

- Follow the principle of least privilege access and run containers with the minimum required permissions. For example, run containers as nonroot unless root privileges are necessary. This makes it difficult for the attackers to escape the container and provides protection from vulnerabilities like *CVE-2020-15257 (https://oreil.ly/txXJL)*, where the root user was able to escape the container.
- Do not use tags for Docker images; rather, pin the base image version in the Dockerfile (i.e., ubuntu:20.08). Mutable tags like *latest* or *master* are updated constantly with features and fixes, which can cause issues while scanning images as a part of your CI/CD pipeline. Additionally, they can cause stability issues for an application (where the underlying dependant library is updated/removed or changed).
- Compress Docker image layers into one single layer. Container images built with tools like Docker or buildah often have multiple layers. These layers show a development history and sometimes end up leaking sensitive information. The best way to compress existing layers is to use a multistage build; there is also an experimental Docker feature (i.e., option `--squash` (*https://oreil.ly/U5RaL*) available in Docker API 1.25+).
- Use container image signing to trust the image. Natively, Kubernetes doesn't have container image verification. Docker Notary can be used to sign images, and by using the Kubernetes admission controller, it is possible to verify an image signature and determine if an image was tampered with by a malicious actor (e.g., if the image is changed while it is seated at the registry).

In the next section, we will review container image scanning.

Container Image Scanning Solution

Container image scanning tools examine the container filesystem to get the metadata to know if there are vulnerable components present in the image. There are many open source and commercial enterprise solutions available in the market that you can use for this purpose. They come with CI/CD integrations and a rich set of scanning features. The solution you choose should answer some of these basic questions:

- Can the image scanner scan OS packages present in a container image for your selected base image?
- Can it scan your application dependencies (does it understand languages used by your application, e.g., Go, Python, Node.js)?
- Can the image scanner detect sensitive files present in the filesystem (certificates, passwords)?
- What is the false positive rate?
- Can it scan binaries (.elf or .exe)?

- What data will be collected by the scanning solution? Does the scanning solution upload your image to its SaaS service, or does the solution only collect package metadata? It is important to understand this due to the risk of data exposure.
- Where will the collected data be stored? On-prem or cloud SaaS? Please review this and choose the option that works with the guidelines set by your security/compliance teams.
- Does the image scanner have an integration with your CI/CD system?

Most scanners collect metadata from filesystems and try to match it with vulnerability information gathered from sources such as the National Vulnerability Database or private intelligence sources to determine the presence of vulnerability. Please note that you should expect both false positives and negatives as a part of scanning. For images with confirmed vulnerabilities, the application and the security team need to work together to analyze the impact and the risk of the CVE to your operation. Remediation steps involve implementing workarounds and patching the image when an update is available.

Many public cloud providers and container registry service providers offer container scanning services. However, there are limitations to the OS versions they support, and most of them don't scan application dependencies. Here the open source world has more to offer. Notable examples of open source tools that can scan application dependencies are Anchore (*https://oreil.ly/QNdyU*), which lets users define policy, and Trivy (*https://oreil.ly/CYP4B*), which is easy to integrate in CI.

Privacy Concerns

Security vulnerabilities and associated information in your product is highly classified data, and in the wrong hands it could be a big liability to the organization. That's why before you choose any solution, it's good practice to verify what data is being collected by a scanning solution and where that data is stored (e.g., within the enterprise on-premise or in the cloud as a part of an SaaS service). If you are buying a commercial solution, it is important to check the contract to know the clauses for damages in case of data breach. Often these clauses can help you understand how serious an organization is about data security. In case you are using an open source solution, please review the documentation to understand the risks of data leakage.

Container Threat Analysis

In addition to traditional image scanning, the area of container threat analysis using sandbox-based solutions is gaining popularity. It is a relatively new area, and we recommend you watch it. These sandbox-based solutions can run Docker images and monitor for container system calls, processes, memory, network traffic (HTTP, DNS, SSL, TCP), and overall behavior of the container, using machine learning and other techniques to determine any malicious activity. Additionally, they can scan container filesystems to check for vulnerabilities as well as malicious binaries. These can be used to detect advanced persistent threats (APTs) and malware.

CI/CD

In this section we will cover various strategies to integrate image scanning solutions into your CI/CD pipeline, best practices to secure your CI/CD pipelines, and techniques to implement organizational policies for CI/CD and vulnerability scanning.

Continuous integration (CI) is a development practice in which each developer's check-in is verified by an automated build, allowing teams to detect problems early. And continuous deployment (CD) is the extension of CI where changes are released for downstream consumers once they pass all release checks.

Integrating security at each step of the development and release process is the goal of CI/CD, which is an integral part of shift-left strategy in DevOps processes. By integrating image scanning into your CI/CD pipeline (see Figure 3-1), development teams can have a verdict available as soon as a check-in from a developer is committed to the repository. The main advantage of this approach is that new security vulnerabilities or threats can be detected at build time. Once an issue is found, all stakeholders and DevOps teams can be notified, typically by failing the CI job. Then respective teams can start working on remediation immediately.

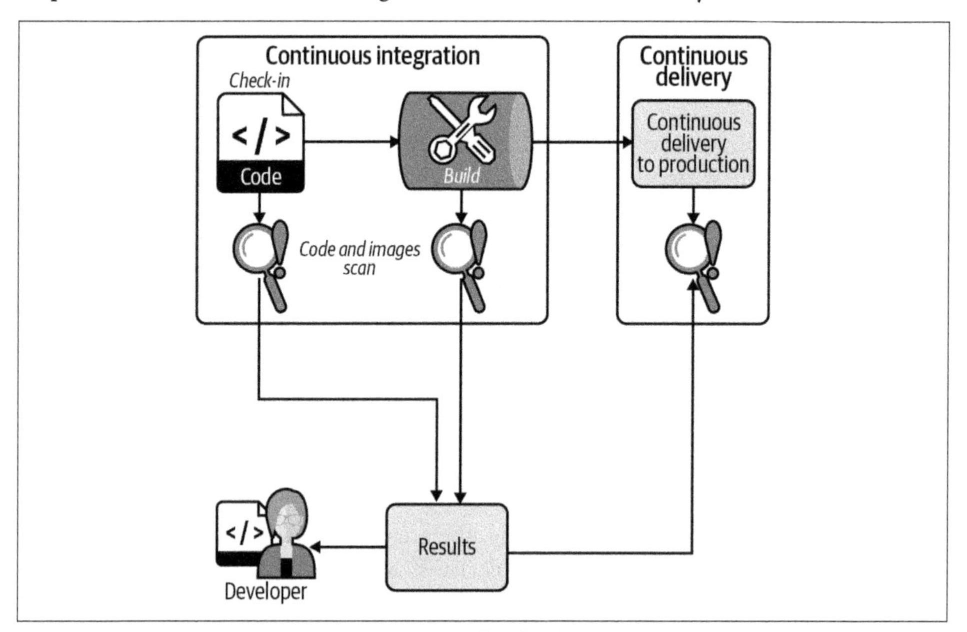

Figure 3-1. Integrating image scanning into the CI/CD process

As shown in Figure 3-1, the scanning should be integrated at each step of the development cycle, from the developer check-in to continuous delivery to production. The rest of this section will be focused on the CI part of the integration to show the granularity of the image scanning process. The following strategies can be applied to each step shown in Figure 3-1 (i.e., code, build [CI], and CD pipeline).

The choice of the specific CI/CD build infrastructure is secondary and can be chosen according to wider requirements. Popular CI/CD providers include but are not limited to Jenkins (*https://oreil.ly/ge9Q1*), Semaphore (*https://oreil.ly/xpiFH*), and CircleCI (*https://circleci.com*). There are four main ways to integrate image scanning into your CI/CD pipeline and build infrastructure, as shown in Figure 3-2.

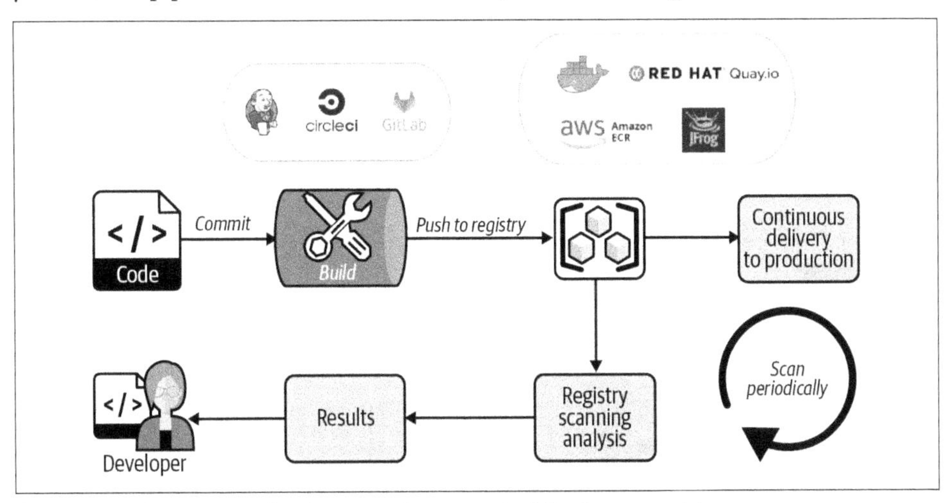

Figure 3-2. Image scanning integrated as a part of the image registry

Scan Images by Registry Scanning Services

As illustrated in Figure 3-2, in this approach as soon as a check-in is available from the developer, CI builds and pushes images to the registry. Then images are periodically scanned by the services integrated within the registry. There are many downsides to this approach; registry providers usually limit themselves to scanning the operating system package layer (e.g., GCR and Quay). Since they show limited information, using vulnerability whitelisting, classification, and tracking timelines to fix various issues can be really cumbersome. Also, most of the time there is no option to write policy tailored to your organization's needs.

When the image scanner finds an issue, the image may have already been consumed or delivered using CD. The registry doesn't keep track of which consumed images consumed and who deployed them. Users using those images may be at risk of compromise and not even know about the vulnerabilities identified by the registry scanning service. Vulnerability remediation only comes after notification, which can be an alert or email to the development team and other stakeholders. It can be anything from a simple update from the upstream provider to complex code and a configuration fix. This kind of remediation effort tends to lag behind the development process.

Figure 3-3 shows how images are built and scanned as a part of the CI process, but images are pushed to the internal registry without considering a verdict from the scanner. If you are an agile DevOps team, you may be building and pushing images to your internal registry every day or every hour as soon as a check-in is available on multiple branches. In this case, you don't want every build to be failing as a part of CI if there is a vulnerability present in the image. Rather, you would get a notification for a vulnerability, and the developer can fix that vulnerability before CD kicks in.

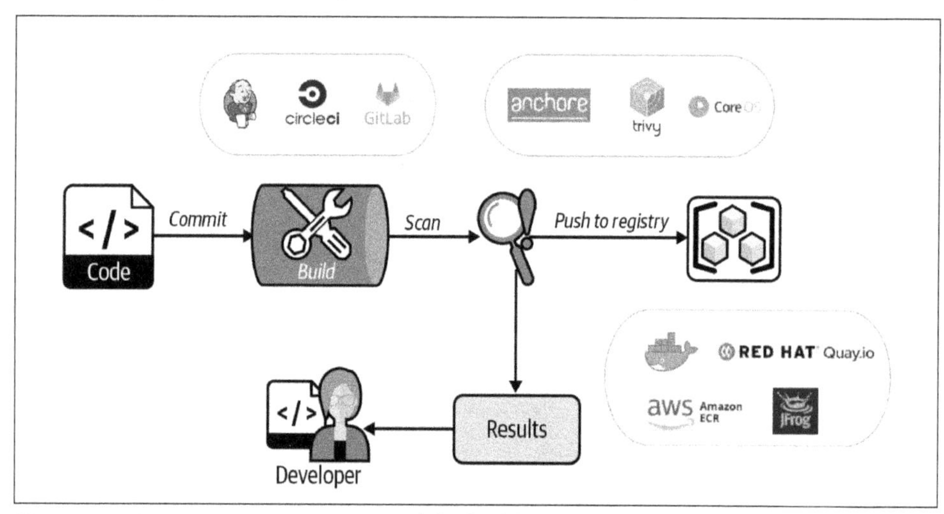

Figure 3-3. Image scanning as a part of the build process

Scan Images After Builds

The drawback of the approach is that even though every image is scanned right after it is built, there is no mechanism to force developers to fix found issues immediately. Additionally, images are available for internal consumption from the registry even if vulnerability is detected, and therefore it could be a weak link in the organization, which can lead to the compromise.

As illustrated in Figure 3-4, a CI job kicks in with a check-in from the developer, which builds and subsequently scans the images. Once the scan is complete, a verdict from the image scanner is evaluated. If it passes, then images are pushed to the registry for internal consumption. If it fails, the developer needs to remediate the issue immediately, as they won't have the latest build available with their changes.

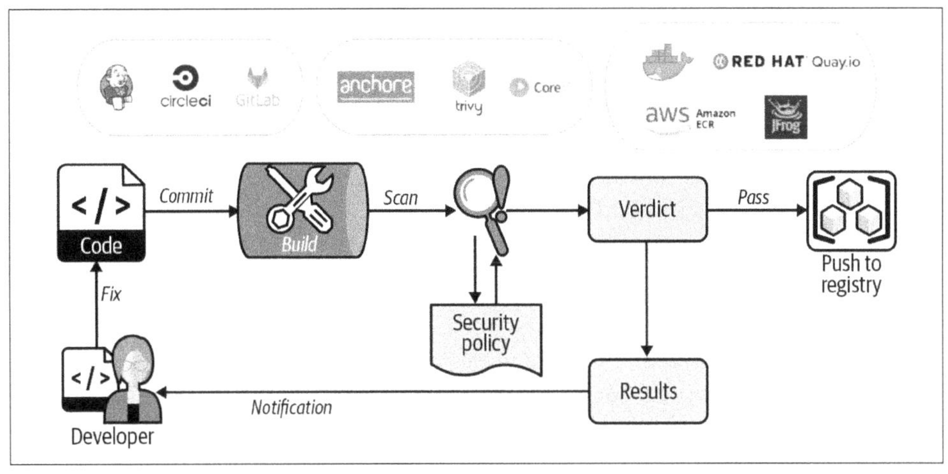

Figure 3-4. Image scanning inline as a part of the CI/CD process

Inline Image Scanning

This pipeline can be difficult to manage initially, depending on the size and velocity of your organization, but once it is mastered, the organization gets much better control over its security posture. The same pipeline design can be utilized in your CD jobs so that your applications are the most secure at the time of release/deployment.

Kubernetes admission controllers can intercept pod creation requests from the Kubernetes API (see Figure 3-5). This mechanism can be used as a last-minute check by triggering a CI job to scan an image that is being deployed on the cluster. Depending on the verdict of the scan, the admission controller can admit or kick out the pod.

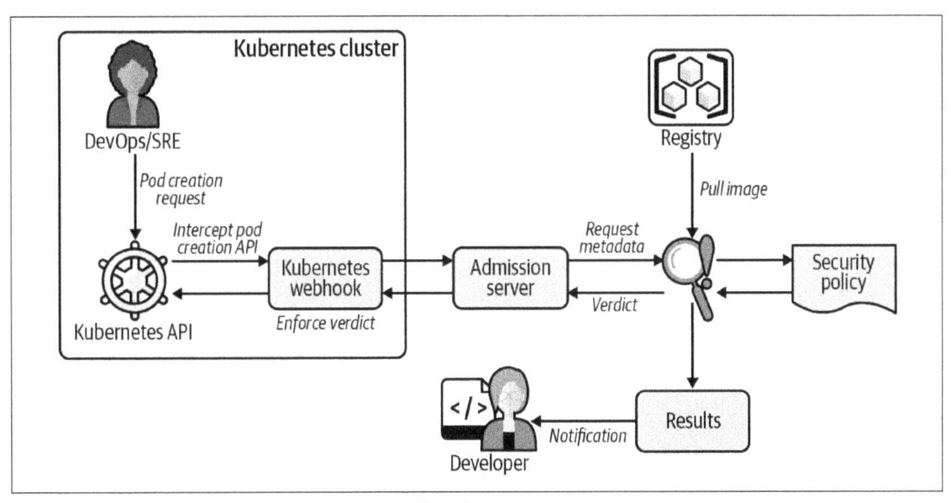

Figure 3-5. Image scanning as a part of pod creation

Kubernetes Admission Controller

This method usually needs a custom or off-the-shelf admission controller that is able to talk to admission servers that respond with a verdict for the scan performed. From a fault tolerance point of view, it's worth noting that if the admission server fails for some reason, then it can impact pod creation in the entire cluster. Whether you want the admission controller to "fail open" (i.e., go ahead with pod creation) or "fail close" (i.e., prevent all pod creation) in the event of admission server failure is the organization's decision; weigh the security risks against the fault tolerance risks.

Since admission controllers are a last-minute check, usually development teams are unaware of the found security vulnerabilities until they look at what is being scanned and rejected at the moment. So it is recommended that this approach be used in combination with earlier methods as a part of your defense strategy.

Securing the CI/CD Pipeline

CI/CD pipelines' autonomous nature and minimum human interaction makes CI/CD an attractive target for attackers. Additionally, the development environment can be overly permissive with a minimum focus on security. Following are the best practices to secure your CI/CD pipelines.

Zero-trust policy for CI/CD environment

Every connection to and from your CI/CD pipeline needs to be scrutinized with a zero-trust policy in place for underlying hosts, infrastructure, and any supporting processes according to your threat model. This will ensure that egress and ingress access to the CI/CD pipeline is managed through a secure policy.

Secure secrets

Review each secret required by your CI/CD pipeline and make sure passwords, access tokens, and encryption keys are called only when required. The design of secrets management needs to consider fine-grained access to secrets, secret usage and changelog capabilities, automated secret rotation, and deactivation and deprecation. We will discuss more about secrets management in the upcoming section of this chapter to help you choose the right strategy.

Access control

Tight access control to CI/CD resources and the separation of user responsibilities is the key to a secure CI/CD pipeline, whether you have a role-based, time-based, or task-based approach. Access control needs to segment the access to the pipeline so that in case of compromise, the blast radius is reduced significantly. Also, use a strong authentication mechanism with two-factor authentication enabled by default.

Audit and monitoring

Access to CI/CD resources needs continuous auditing and monitoring to determine excessive access, access deprecation in case a user leaves the organization or changes job roles, abuse, or suspicious user behavior.

Organization Policy

To tackle challenges presented by CI/CD pipelines and scanning, a global organization policy is required. The policy needs to call out access requirements to CI/CD resources, separation of user responsibilities, secret management, logging and monitoring requirements, and audit policy.

Vulnerability scans can overwhelm teams initially. Hence, development, DevOps, and security teams need clear directives on product vulnerability discovery and assessment, risk, remediation, and timelines to close the issues based on the organization's threat model.

Vulnerability scanning solutions can have codified policies where images can be admitted or rejected based on scan results and the risk tolerance of the organization.

The process to build an effective policy needs to be iterative and based on continuous feedback to achieve a tailored policy, balancing security and performance according to your industry, size, workflows, and compliance requirements.

Secrets Management

A secret can be anything that is used to authenticate and authorize users, groups, or entities. It can be a username/passwords, API token, or TLS certificate. When an application or microservice is moved to Kubernetes, the early design choice developers need to make is where to store these secrets and how to retrieve them and make them available in an application as needed without compromising the security posture of the application. Following are the top methods to achieve this objective.

etcd to Store Secrets

A common scenario for an application moving to Kubernetes is to store a secret in a Base64-encoded format in etcd as a key-value pair. Etcd is a supported datastore in Kubernetes deployments. These secrets can be made available inside the container as a volume mount or an environment variable from within Kubernetes deployment specs. Since environment variables are stored in memory, it's hard to extract secrets, compared with volume mounts, which store secrets on a container filesystem. The access to etcd is backed by Kubernetes RBAC, which brings needed security and flexibility.

Etcd provides strong concurrency primitives, linearizable reads, and APIs to manage secrets at scale. The downside of this approach is secrets are stored in plain text (Base64 encoding) and are retrieved and sent in plain text unless etcd is configured to encrypt communication using TLS. In Chapter 2, you learned the strategy to encrypt data at rest, where secrets can be encrypted while they are stored in etcd.

Additionally, secrets stored in etcd are not versioned or recoverable once deleted, and access to etcd is not audited, so anyone who has access to etcd can access all secrets. Since etcd is a Kubernetes datastore, the broader secrets management requirements of the organization are not fulfilled.

Secrets Management Service

To solve the organization's encryption and secrets management requirement, secrets management services can be utilized from cloud providers. All major public cloud providers provide secrets management services.

The most popular cloud providers' secrets management services are AWS Secrets Manager (*https://oreil.ly/7tuNL*), Google Secret Manager (*https://oreil.ly/JnfOA*), and Azure Key Vault (*https://oreil.ly/MaMgd*).

A notable third-party example is HashiCorp Vault (*https://oreil.ly/xDMd3*), which can be used as a centralized secrets manager. It provides many features to fulfill an organization's end-to-end secrets management requirements (key management, encryption, rotation, PKI, storage, replication, revocation, logging, monitoring, audit, etc.). This tool can be used in conjunction with a cloud provider's secrets management services—for example, when the vault is initialized, initial keys can be encrypted and stored into Cloud KMS so that the operator won't have to handle plain-text keys.

Kubernetes Secrets Store CSI Driver

The secret store Container Storage Interface (CSI) driver (*https://oreil.ly/kOio7*) integrates external secret stores like Azure, GCP, AWS, and Vault from HashiCorp into Kubernetes using CSI, which is generally available since version 1.13.

In a nutshell, the CSI driver authenticates with your secret store service using volume attributes and mounts needed secrets into the pod seamlessly. This approach avoids the use of the Kubernetes etcd datastore and allows you to scale and manage the organization's secrets effectively.

Secrets Management Best Practices

Following are the best practices to consider when managing secrets in Kubernetes.

Avoid secrets sprawl

The main motive of secrets management is to avoid secrets sprawl, where your application secrets are spread across places like config files, yamls, and Git repositories. This is usually a sign of the lack of a secrets management workflow in the organization. The only way to mitigate secrets sprawl is to have a centralized secrets management strategy in place, one where credentials can be stored and retrieved securely from a single point and used by the entire organization with proper authorization, logging, and monitoring mechanisms in place.

Use anti-affinity rules

Ideally, a secrets management solution should be a single process on a small number of dedicated VMs or dedicated hosts. Since you may need to run this solution on Kubernetes as a microservice, it will be a process running in a dedicated pod. But the issue becomes on which nodes these pods should be running. Here anti-affinity helps by distributing pods on required nodes, which are classified to run a secrets management solution.

Data encryption (transit and rest)

By default, Kubernetes insecurely stores and transmits secrets. It is paramount to configure or have a solution that can use end-to-end TLS encryption where secrets are encrypted in transit. At the same time, have a mechanism in place to store secrets in encrypted form. See Chapter 2 for options on how to achieve this.

Use automated secret rotation

Organizations follow different time frames for different secrets for rotation, but with the advent of automated secret rotation, you can do it daily or even on an hourly basis. Cloud secrets management services and external third-party solutions both can help to rotate and manage secrets in an automated fashion.

Ephemeral or dynamic secret

Ephemeral or dynamic secrets are temporary, on-demand generated secrets that typically have a short time to live and are destroyed after that time interval. These secrets can be made available to the class of the application or to the operations team as needed. If a secret is discovered by an attacker (for example, if leaked via debug logs, application code, or accidentally exposed via GitHub), the secrets would have been changed in a short window of time, protecting applications. Additionally, they can help trace the attacker's footsteps in the infrastructure, since it becomes easy to determine the time frame in which the secret was discovered by the attacker. HashiCorp Vault and CyberArk Conjur are some of the third-party secrets providers that provide such features.

Enable audit log

Having an audit log in your secrets management solution provides visibility into secrets and their uses by the organization. An audit log can be critical for determining intentional or unintentional compromises, the blast radius of the attack, and related forensic steps.

Store secrets in container memory

When a containerized application receives a secret, don't store the secret on disk (or in volumeMount available in the host). Rather, store it in memory so that in case of a compromise, those secrets are not easily available to the attacker.

Secret zero problem

Many secrets management solutions follow envelope encryption where DEKs are protected by a KEK. The KEK is considered secret zero. If an attacker compromises the KEK, then they can decrypt the DEK and subsequently the data encrypted by the DEK. The combination of cloud providers' IAM and KMS can be used to help protect secret zero. (Though, of course, these in turn effectively have their own secret zero further up the trust chain, which you must treat as highly sensitive.)

Use your Certificate Authority

As part of defense-in-depth, end-to-end TLS implementation can be done using a custom Certificate Authority (CA). Here an organization can choose to sign its certificate using its own CA. In this case, the service can only be accessed by presenting a certificate signed by the organization.

Authentication

Once you are ready with your hardened images, CI/CD pipeline, and secrets management strategy, it's time for Kubernetes' authentication and authorization strategy. Kubernetes allows numerous authentication mechanisms; in simplest form; authentication is done with certificates, tokens, or basic authentication (username and password). Additionally, webhooks can be used to verify bearer tokens, and external OpenID providers can be integrated.

Let's take a closer look at each authentication method available in Kubernetes. Configuration for authentication methods is out of scope for this book.

X509 Client Certificates

There are two ways to sign a client certificate so that it can be used to authenticate with the Kubernetes API. First is internally signing the certificate using the Kubernetes API. It involves the creation of a certificate signing request (CSR) by a client.

Administrators can approve or deny the CSR. Once approved, the administrator can extract and provide a signed certificate to the requesting client or user. This method cannot be scaled for large organizations as it requires manual intervention.

The second method is to use enterprise PKI, which can sign the client-submitted CSR. Additionally, the signing authority can send signed certificates back to clients. This approach requires the private key to be managed by an external solution.

Bearer Token

Kubernetes service accounts use bearer tokens to authenticate with Kubernetes API. The simplest way to use a bearer token is to create a new service account. Kubernetes API automatically issues a random token associated with the service account, which can be retrieved and used to authenticate that account.

Bearer tokens can be verified using a webhook, which involves API configuration with option `--authentication-token-webhook-config-file`, which includes the details of the remote webhook service.

Kubernetes internally uses Bootstrap and Node authentication tokens to initialize the cluster. Also, there''s a less secure option available using a static token file that can be provided using the `--token-auth-file` option while configuring Kubernetes API.

OIDC Tokens

The OpenID Connect protocol (*https://oreil.ly/8t476*) is built by expanding the existing OAuth2 protocol. Kubernetes does not provide an OpenID Connect identity provider. You can use identity providers like Google or Azure or run your own identity provider using dex (*https://oreil.ly/Rd2sg*), keycloak (*https://oreil.ly/ah3IY*), or UAA (*https://oreil.ly/oMhzC*). These external identity providers can easily be integrated with your authentication workflows as required, as well as support native identity provider capabilities (e.g., enterprise using lightweight directory access protocol).

Authentication Proxy

A proxy can be used to establish a trust connection with Kubernetes API. Kubernetes API can identify users from request headers such as X-Remote-User, which is set by authentication proxy as it authenticates users on behalf of the Kubernetes API. Authentication proxy can authenticate users as needed according to your workflow.

Anonymous Requests

If a request to Kubernetes API is not rejected by any configured authentication method, then it is treated as an anonymous request (i.e., a request without a bearer token).

It is important to note that for Kubernetes version 1.6 and above, anonymous access is enabled by default for an authorization mode other than `AlwaysAllow`. It can be disabled by adding the option `--anonymous-auth=false` while configuring Kubernetes API.

User Impersonation

This is a subtle authentication mechanism where a user with certain access to Kubernetes can impersonate another user by setting additional headers in the request to Kubernetes API with the details of the impersonated user.

This mechanism allows Kubernetes API to process requests as per the impersonated user's privileges and context. Additionally, Kubernetes API can log who has impersonated whom and other relevant details from a request as necessary, which can be useful during monitoring and audit.

Authorization

In this section we will cover available authorization methods, RBAC in Kubernetes, namespaced RBAC, and caveats.

Kubernetes has multiple authorization mechanisms such as Node, ABAC, RBAC, and AlwaysDeny/AlwaysAllow, though RBAC is the industry standard in Kubernetes.

Node

Node authorization is used by Kubernetes internally. It's a special-purpose authorization mode that specifically authorizes API requests made by kubelets. It enables read, write, and auth-related operations by kubelet. In order to successfully make a request, kubelet must use a credential that identifies it as being in the system:nodes group.

ABAC

Kubernetes defines attribute-based access control (ABAC) as "an access control paradigm whereby access rights are granted to users through the use of policies which combine attributes together." ABAC can be enabled by providing a *.json* file to `--authorization-policy-file` and `--authorization-mode=ABAC` options in Kubernetes API configurations. The *.json* file needs to be present before Kubernetes API can be invoked.

AlwaysDeny/AlwaysAllow

The AlwaysDeny or AlwaysAllow authorization mode is usually used in development environments where all requests to the Kubernetes API need to be allowed or denied. AlwaysDeny or AlwaysAllow mode can be enabled using option

`--authorization-mode=AlwaysDeny/AlwaysAllow` while configuring Kubernetes API. This mode is considered insecure and hence is not recommended in production environments.

RBAC

Role-based access control is the most secure and recommended authorization mechanism in Kubernetes. It is an approach to restrict system access based on the roles of users within your cluster. It allows organizations to enforce the principle of least privileges. Kubernetes RBAC follows a declarative nature with clear permissions (operations), API objects (resources), and subjects (users, groups, or ServiceAccounts) declared in authorization requests. Applying an RBAC in Kubernetes is a two-step process. First is to create a Role or ClusterRole. The latter is a global object where the former is a namespace object. A Role or ClusterRole is made up of verbs, resources, and subjects, which provide a capability (verb) on a resource, as shown in Figure 3-6. The second step is to create a ClusterRoleBinding where the privileges defined in step 1 are assigned to the user or group.

Let's take an example where a dev-admins group needs to have read access to all the secrets in the cluster. Step 1 is to create a ClusterRole secret-reader, which allows the reading of the secrets via various operations (get, list), and step 2 is binding it to a subject (i.e., users, groups, or ServiceAccounts) to provide access. In this case the group dev-admins allows group users to read secrets globally.

Figure 3-6 is an example of how you can create a ClusterRole that allows you to define access to resources. The example on the right shows how you can bind the ClusterRole to a group of users.

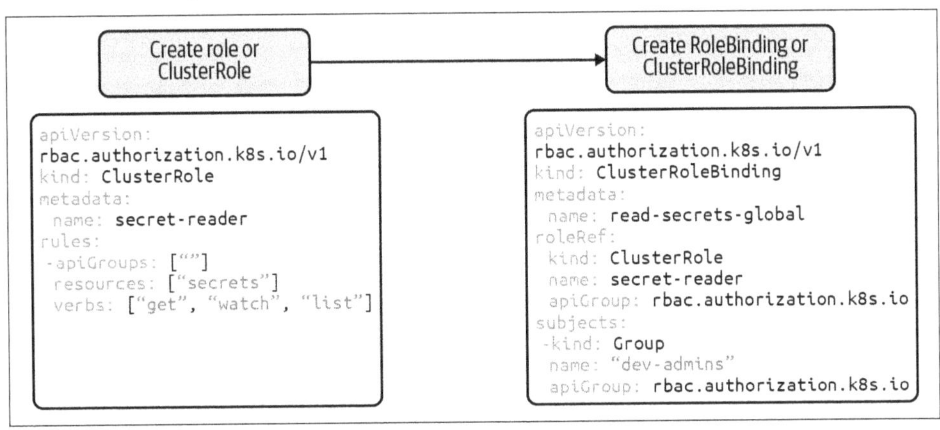

Figure 3-6. Kubernetes role and role binding

Figure 3-7 shows an overview of RBAC; you can use any combination of operation, resource, or subject as required.

Apart from these resources, there are Kubernetes nonresources like /healths or /version APIs that can be controlled using RBAC as well if needed.

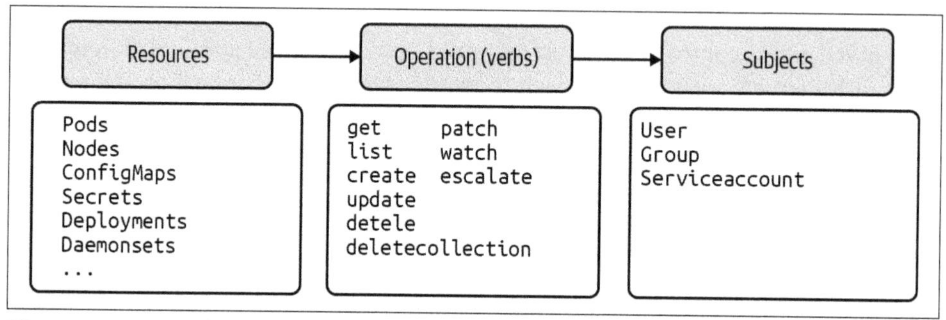

Figure 3-7. RBAC for Kubernetes resources

Namespaced RBAC

In earlier examples you saw an RBAC that was applied globally in the cluster. It is possible to apply a similar RBAC to namespaces where resources within a namespace can be subject to this RBAC policy. The namespaced resources Role and RoleBinding should be used for configuring a namespace policy.

There are few caveats you should be aware of while using namespaced RBAC:

- Roles and RoleBindings are the only namespaced resources.
- ClusterRoleBindings (global resource) cannot be used with Roles, which is a namespaced resource.
- RoleBindings (namespaced resource) cannot be used with ClusterRoles, which are global resources.
- Only ClusterRoles can be aggregated.

Privilege Escalation Mitigation

Kubernetes RBAC reduces an attacker's ability to escalate their own privileges by editing roles or role bindings. This behavior is enforced at the API level in Kubernetes and even applies when the RBAC authorizer is not in use:

- If user user-no-secret doesn't have the ability to list secrets cluster-wide, they cannot create a ClusterRole or ClusterRoleBinding containing that permission.
- For user user-no-secret to get a list of secret privileges, they will need one of the following:

- Grant them a role that allows them to create/update Roles, ClusterRoles, RoleBindinsg, or ClusterRoleBindings.
- Provide them explicit permission with the verb escalate on these resources.

Conclusion

In this chapter we covered the following key concepts that will help you understand security tools and best practices to build and deploy workloads:

- It is insufficient to use available base images from Docker as they are for your containers; you must spend time ensuring that your container images are hardened and built with security in mind. Just like in software development, finding vulnerabilities at build time is far cheaper than finding vulnerabilities after software is deployed.
- There are several ways to add image scanning to your CI/CD process. We explored various well-known approaches like registry scan, build time or inline scan, and using Kubernetes admission controller to help you add image scanning to your CI/CD pipelines. We also looked at securing CI/CD pipelines and adding organization policy to effectively craft a workflow for your organization.
- We covered the approaches and best practices to secret management.
- Finally, we covered available Kubernetes authentication and authorization mechanisms. We recommend you use RBAC to mitigate privilege escalation.

CHAPTER 4

Workload Runtime Security

With contributions from Manoj Ahuje,
Senior Threat Intelligence Research Engineer at Tigera

Kubernetes' default pod provisioning mechanism has a wide attack surface that can be used by adversaries to exploit the cluster or escape the container. In this chapter you will learn how to implement pod security policies (PSPs) to limit the attack surface of the pods and how to monitor processes (e.g., process privileges), file access, and runtime security for your workloads. Here are a few specifics of what we will discuss:

- We will cover the implementation details of PSPs, like pod security contexts, and also explain the limitations of PSPs. Note PSPs are deprecated as of Kubernetes v1.21; however, we will cover this topic in this chapter as we are aware that PSPs are widely used.
- We will discuss process monitoring, which focuses on the need for Kubernetes-native monitoring to detect suspicious activities. We will cover runtime monitoring and enforcement using kernel security features like seccomp, SELinux, and AppArmor to prevent containers from accessing host resources.
- We will cover both detection and runtime defense against vulnerabilities, workload isolation, and a blast radius containment.

Pod Security Policies

Kubernetes provides a way to securely onboard your pods and containers by using PSPs. They are a cluster-scoped resource that checks for a set of conditions before a pod is admitted and scheduled to run in a cluster. This is achieved via a Kubernetes admission controller, which evaluates every pod creation request for compliance with the PSP assigned to the pod.

Please note that PSPs are deprecated with Kubernetes release 1.21 and are scheduled to be removed in release 1.25. They are widely used in production clusters, though, and therefore this section will help you understand how they work and what best practices are for implementing PSPs.

PSPs let you enforce rules with controls like *pods should not run as root* or *pods should not use host network, host namespace, or run as privileged.* The policies are enforced at pod creation time. By using PSPs you can make sure pods are created with the minimum privileges needed for operation, which reduces the attack surface for your application. Additionally, this mechanism helps you to be compliant with various standards like PCI, SOC 2, or HIPAA, which mandates the use of principle of least privilege access. As the name suggests, the principle requires that any process, user, or, in our case, workload be granted the least amount of privileges necessary for it to function.

Using Pod Security Policies

Kubernetes PSPs are recommended but implemented via an optional admission controller. The enforcement of PSPs can be turned on by enabling an admission controller. That means the Kubernetes API server manifest should have a PodSecurityPolicy plug-in in its --enable-admission-plugins list. Many Kubernetes distros do not support or by default disable PSPs, so it's worth checking while choosing the Kubernetes distros.

Once the PSPs are enabled, it's a three-step process to apply PSPs, as shown in Figure 4-1. A best practice is to apply PSPs to groups rather than individual service accounts.

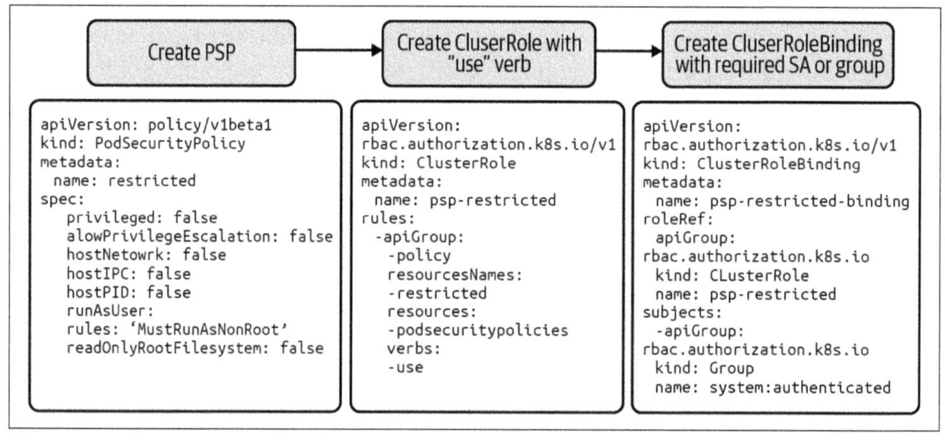

Figure 4-1. Process to apply PSPs

Step 1 is to create a PSP. Step 2 is to create ClusterRole with the `use` verb, which authorizes pod deployment controllers to use the policies. Then step 3 is to create ClusterRoleBindings, which is used to enforce policy for the groups (i.e., system:authenticated or system:unauthenticated) or service accounts.

A good starting point is the PSP template from the Kubernetes project:

```
apiVersion: policy/v1beta1
kind: PodSecurityPolicy
metadata:
  name: restricted
  annotations:
    seccomp.security.alpha.kubernetes.io/allowedProfileNames: |
    'docker/default,runtime/default'
    apparmor.security.beta.kubernetes.io/allowedProfileNames: 'runtime/default'
    seccomp.security.alpha.kubernetes.io/defaultProfileName:  'runtime/default'
    apparmor.security.beta.kubernetes.io/defaultProfileName:  'runtime/default'
spec:
  privileged: false
  # Required to prevent escalations to root.
  allowPrivilegeEscalation: false
  # This is redundant with non-root + disallow privilege escalation,
  # but we can provide it for defense in depth.
  requiredDropCapabilities:
    - ALL
  # Allow core volume types.
  volumes:
    - 'configMap'
    - 'emptyDir'
    - 'projected'
    - 'secret'
    - 'downwardAPI'
    # Assume that persistentVolumes set up by the cluster admin are safe to use.
    - 'persistentVolumeClaim'
  hostNetwork: false
  hostIPC: false
  hostPID: false
  runAsUser:
    # Require the container to run without root privileges.
    rule: 'MustRunAsNonRoot'
  seLinux:
    # This policy assumes the nodes are using AppArmor rather than SELinux.
    rule: 'RunAsAny'
  supplementalGroups:
    rule: 'MustRunAs'
    ranges:
      # Forbid adding the root group.
      - min: 1
        max: 65535
  fsGroup:
    rule: 'MustRunAs'
    ranges:
```

```
    # Forbid adding the root group.
    - min: 1
      max: 65535
  readOnlyRootFilesystem: false
```

In the following example, you apply this policy to all authenticated users using Kubernetes role-based access control:

```
apiVersion: rbac.authorization.k8s.io/v1
kind: ClusterRole  Policy
metadata:
  name: psp-restricted
rules:
- apiGroups:
  - policy
  resourceNames:
  - restricted
  resources:
  - podsecuritypolicies
  verbs:
  - use
---
apiVersion: rbac.authorization.k8s.io/v1
kind: ClusterRoleBinding
metadata:
  name: psp-restricted-binding
roleRef:
  apiGroup: rbac.authorization.k8s.io
  kind: ClusterRole
  name: psp-restricted
subjects:
  - apiGroup: rbac.authorization.k8s.io
    kind : Group
    name: system:authenticated
```

Pod Security Policy Capabilities

Let's focus on the capabilities provided by PSPs that you can utilize as required by your use case and internal threat model. You can follow the example PSP template we just discussed to build your own PSPs. In this template most of the PSP capabilities are utilized to formulate a restrictive policy.

To explain the impact of a capability, let's take a look at an example where you see capabilities granted to the pod created with privileged:true and with privileged:false. A Linux utility capsh (*https://oreil.ly/UuDcu*) can be used to evaluate the permissions of containerized root users. As you can see in Figure 4-2, the privileged pod has a plethora of capabilities in its Linux namespace, which translates to a wider attack surface for an attacker to escape your container.

```
root@attacker-normal:/# id
uid=0(root) gid=0(root) groups=0(root)
root@attacker-normal:/# capsh --print
Current: cap_chown,cap_dac_override,cap_fowner,cap_fsetid,cap_kill,cap_setgid,cap_setuid,cap_setpcap,cap
_net_bind_service,cap_net_raw,cap_sys_chroot,cap_mknod,cap_audit_write,cap_setfcap=eip
```

```
root@attacker-priviledged:/# capsh --print
Current: =eip
Bounding set =cap_chown,cap_dac_override,cap_dac_read_search,cap_fowner,cap_fsetid,cap_kill,cap_setgid,c
ap_setuid,cap_setpcap,cap_linux_immutable,cap_net_bind_service,cap_net_broadcast,cap_net_admin,cap_net_r
aw,cap_ipc_lock,cap_ipc_owner,cap_sys_module,cap_sys_rawio,cap_sys_chroot,cap_sys_ptrace,cap_sys_pacct,c
ap_sys_admin,cap_sys_boot,cap_sys_nice,cap_sys_resource,cap_sys_time,cap_sys_tty_config,cap_mknod,cap_le
ase,cap_audit_write,cap_audit_control,cap_setfcap,cap_mac_override,cap_mac_admin,cap_syslog,cap_wake_ala
rm,cap_block_suspend,cap_audit_read
```

Figure 4-2. Pod capabilities for default and privileged pods

Table 4-1 summarizes the capabilities for pods as described in the Kubernetes PSP documentation (*https://oreil.ly/FSDGN*).

Table 4-1. Pod capabilities

Field	Uses
privileged	Allow containers to gain capabilities that include access to host mounts, filesystem to change settings, and many more. You can check capabilities with command capsh --print.
hostPID, hostIPC	Give container access to host namespaces where process and Ethernet interfaces are visible to it.
hostNetwork, hostPorts	Give container IP access to the host network and ports.
volumes	Allow volumes types like configMap, emtyDir, or secret.
allowedHostPaths	Allow the whitelisting of host paths that can be used by hostPath volumes (i.e., /tmp).
allowedFlexVolumes	Allow specific FlexVolume drivers (i.e., azure/kv).
fsGroup	Set a GID or range of GID that owns the pod's volumes.
readOnlyRootFilesystem	Set the container's root filesystem to read-only.
runAsUser, runAsGroup, supplementalGroups	Define containers UID and GID. Here you can specify non-root user or groups.
allowPrivilegeEscalation, defaultAllowPrivilegeEscalation	Restrict privilege escalation by process.
defaultAddCapabilities, requiredDropCapabilities, allowedCapabilities	Add or drop Linux capabilities (*https://oreil.ly/H87Jc*) as needed.
SELinux	Define the SELinux context of the container.
allowedProcMountTypes	Allowed proc mount types by container.
forbiddenSysctls,allowedUnsafeSysctls	Set the sysctl profile used by the container.
annotations	Set the AppArmor and seccomp profiles used by containers.

AppArmor and seccomp profiles are used with PSP annotation where you can use the runtime's (Docker, CRI) default profile or choose a custom profile loaded on the host by you. You will see more about these defenses in "Process Monitoring" on page 59.

Pod Security Context

Unlike PSPs, which are defined cluster-wide, a pod securityContext can be defined at runtime while creating a deployment or pod. Here is a simple example of pod securityContext in action, where the pod is created with the root user (`uid=0`) and allows only four capabilities:

```
kind: Pod
apiVersion: v1
metadata:
  name: attacker-privileged-test
  namespace: default
  labels:
    app: normal-app
spec:
  containers:
  - name: attacker-container
    image: alpine:latest
    args: ["sleep", "10000"]
    securityContext:
      runAsUser: 0
      capabilities:
        drop:
          - all
        add:
          - SYS_CHROOT
          - NET_BIND_SERVICE
          - SETGID
          - SETUID
```

This code snippet shows how you can create a pod running a root but limited to a subset of capabilities by specifying a security context. Figure 4-3 shows commands you can run to verify that the pod runs as root with the limited set of capabilities.

```
root@attacker-privileged-test:/# id
uid=0(root) gid=0(root) groups=0(root)
root@attacker-privileged-test:/# capsh --print
Current: cap_setgid,cap_setuid,cap_net_bind_service,cap_sys_chroot=eip
```

Figure 4-3. Four allowed pod capabilities

Pod securityContext, as shown in Figure 4-3, can be used without enabling PSPs cluster-wide, but once the PSPs are enabled, you need to define securityContext to make sure pods are created properly. Since the securityContext has a PSP construct, all the PSPs' capabilities apply to securityContext.

Limitations of PSPs

Some of the limitations of PSPs include:

- PodSecurityPolicySpec has references to allowedCapabilities, privileged, or hostNetwork. These enforcements can work only on Linux-based runtimes.
- If you are creating a pod using controllers (e.g., replication controller), it's worth checking if PSPs are authorized for use by those controllers.
- Once PSPs are enabled cluster-wide and a pod doesn't start because of an incorrect PSP, it becomes hectic to troubleshoot the issue. Moreover, if PSPs are enabled cluster-wide in production clusters, you need to test each and every component in your cluster, including dependencies like mutating admission controllers and conflicting verdicts.
- Azure Kubernetes Service (AKS) (*https://oreil.ly/Z99lf*) has deprecated support for PSPs and preferred OPA Gatekeeper for policy enforcement to support more flexible policies using the OPA engine.
- PSP are deprecated and scheduled to be removed by Kubernetes v1.25.
- Kubernetes can have edge cases where PSPs can be bypassed (e.g., TOB-K8S-038 (*https://oreil.ly/PqfNQ*)).

Now that you understand PSPs, best practices to implement them, and the limitations of PSPs, let's look at process monitoring.

Process Monitoring

When you containerize a workload and run it on a host with an orchestrator like Kubernetes, there are a number of layers you need to take into consideration for monitoring a process inside a container. These start with container process logs and artifacts, filesystem access, network connections, system calls required, kernel permission (specialized workload), Kubernetes artifacts, and cloud infrastructure artifacts. Usually your organization's security posture depends on how good your solutions are in stitching together these various log contexts. And this is where the traditional monitoring system fails measurably and a need for Kubernetes' native monitoring and observability arises. Traditional solutions, like endpoint detection and response (EDR) and endpoint protection systems, have the following limitations when used in Kubernetes clusters:

- They are not container aware.
- They are not aware of container networking and typically see activity from the host perspective, which can lead to false negatives on attackers' lateral movements.

- They are blind to traffic between containers and don't have any sight of underlays like IPIP or VXLAN.
- They are not aware of process privileges and file permissions of containers accessing the underlying host.
- They are not aware of the Kubernetes container runtime interface (CRI) or its intricacies and security issues, which can lead to containers being able to access resources on the host. This is also known as *privilege escalation*.

In the following sections, we will go over various techniques you can use for process monitoring. First we look at monitoring using various logs available in Kubernetes; then we explore seccomp, SELinux, and AppArmor features that allow you to control what a process can access (e.g., system calls, filesystem, etc.)

Kubernetes Native Monitoring

As shown in Figure 4-4, each layer leading up to your containerized application process introduces monitoring and logging requirements and a new attack surface that is different from what traditional IT security practitioners are used to for monitoring networks and applications. The challenge is to reduce this monitoring overhead, as it can get really expensive for the storage and compute resources. The topic of metric collection and how to do this efficiently is covered in detail in Chapter 5.

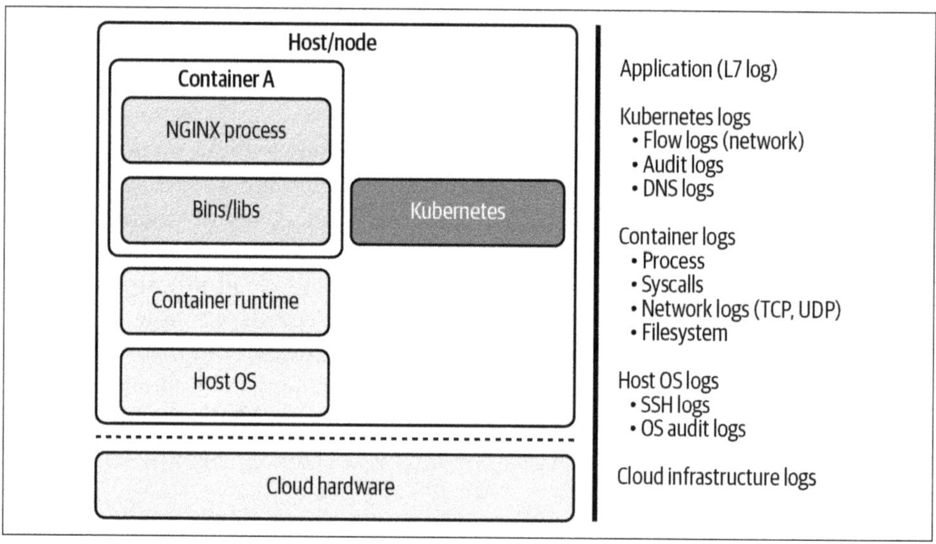

Figure 4-4. Kubernetes native monitoring

In order to build defenses in each layer, the following are some options you should consider incorporating while choosing solutions:

- Ability to block processes spawned by each container or Kubernetes orchestration creating containers.
- Monitor kernel system calls used by each container process and ability to filter, block, and alert on suspicious calls to prevent containers accessing host resources.
- Monitor each network connection (socket) originated by a container process and ability to enforce network policy.
- Ability to isolate a container using network policy (or a node running this container) and pause it to investigate suspicious activities and collect forensics data in Kubernetes. The `pause` command (*https://oreil.ly/LE52U*) for Docker-based containers suspends processes in a container to allow for detailed analysis. Note that pausing a container will cause the container to suspend normal operation and should be used as a response to an event (e.g., security incident).
- Monitor filesystem reads and writes to know filesystem changes (binaries, packages) and additional isolation through mandatory access control (MAC) to prevent privilege escalations.
- Monitor the Kubernetes audit log to know what Kubernetes API requests clients are making and detect suspicious activity.
- Enable a cloud provider's logging for your infrastructure and ability to detect suspicious activity in the cloud provider's infrastructure.

There are many enterprise and open source solutions (e.g., Falco) that target groups of layers using various tools and mechanisms (like ebpf, kprobes, ptrace, tracepoints, etc.) to help build defense at various layers. You should look at their threat model and choose solutions that fulfill their requirements.

In the next section you will see some of the mechanisms that are offered by Kubernetes by bringing Linux defenses closer to the container, which will help you in monitoring and reducing the attack surface at various layers. The previous section focused on monitoring to allow you to detect unintended (malicious) behavior. The following mechanisms allow you to set controls to prevent unintended (malicious) behavior.

Kernel security features like seccomp, AppArmor, and SELinux can control what system calls are required for your containerized application, virtually isolate and customize each container for the workload it is running, and use MAC to provide access to resources like volume or filesystem that prevent container breakouts efficiently. Just using the feature with default settings can tremendously reduce the attack surface throughout your cluster. In the following sections you will look at each defense in depth and how it works in the Kubernetes cluster so that you can choose the best option for your threat model.

Seccomp

Seccomp is a Linux kernel feature that can filter system calls executed by the container on a granular basis. Kubernetes lets you automatically apply seccomp profiles loaded onto a node by Kubernetes runtimes like Docker (*https://www.docker.com*), podman (*https://oreil.ly/O4ZUZ*), or CRI-O (*https://oreil.ly/Cgrep*). A simple seccomp profile consists of a list of syscalls and the appropriate action to take when a syscall is invoked. This action reduces the attack surface to only allowed syscalls, reducing the risk of privilege escalation and container escape.

In the following seccomp profile, a default action is `SCMP_ACT_ERRNO`, which denies a system call. But defaultAction for syscall chmod is overwritten with `SCMP_ACT_ALLOW`. Usually seccomp profiles are loaded into the directory /var/lib/kubelet/seccomp on all nodes by your runtimes. You can add your custom profile at the same place:

```
{
    "defaultAction": "SCMP_ACT_ERRNO",
    "architectures": [
        "SCMP_ARCH_X86_64",
        "SCMP_ARCH_X86",
        "SCMP_ARCH_X32"
    ],
    "syscalls": [
        {
            "names": [
                "chmod",
            ],
            "action": "SCMP_ACT_ALLOW"
        }
    ]
}
```

To find the system calls used by your application, you can use `strace` as shown in the next example. For this example, you can list syscalls used by `curl` utility as follows:

```
$ strace -c -S name curl -sS google.com

% time     seconds  usecs/call     calls    errors syscall
------ ----------- ----------- --------- --------- ----------------
  4.56    0.000242           6        43        43 access
  0.06    0.000003           3         1           arch_prctl
  1.28    0.000068          10         7           brk
  0.28    0.000015          15         1           clone
  4.62    0.000245           5        48           close
  1.38    0.000073          73         1         1 connect
  0.00    0.000000           0         1           execve
  0.36    0.000019          10         2           fcntl
  4.20    0.000223           5        48           fstat
  0.66    0.000035           3        11           futex
  0.23    0.000012          12         1           getpeername
  0.13    0.000007           7         1           getrandom
```

```
  0.19    0.000010         10         1          getsockname
  0.24    0.000013         13         1          getsockopt
  0.15    0.000008          4         2          ioctl
 13.96    0.000741          7       108          mmap
 11.94    0.000634          7        85          mprotect
  0.32    0.000017         17         1          munmap
 11.02    0.000585         13        45        1 openat
  0.11    0.000006          6         1          pipe
 19.50    0.001035        115         9          poll
  0.08    0.000004          4         1          prlimit64
  5.43    0.000288          6        45          read
  0.41    0.000022         22         1          recvfrom
 11.47    0.000609         17        36          rt_sigaction
  0.08    0.000004          4         1          rt_sigprocmask
  1.00    0.000053         53         1          sendto
  0.06    0.000003          3         1          set_robust_list
  0.04    0.000002          2         1          set_tid_address
  2.22    0.000118         30         4          setsockopt
  1.60    0.000085         43         2          socket
  0.08    0.000004          4         1        1 stat
  2.35    0.000125         21         6          write
------ ----------- ----------- --------- --------- ----------------
100.00    0.005308                  518       46 total
```

The default seccomp profiles provided by the Kubernetes runtime contain a list of common syscalls that are used by most of the applications. Just enabling this feature forbids the use of dangerous system calls, which can lead to a kernel exploit and a container escape. The default Docker runtime seccomp profile is available (*https://oreil.ly/pGX2O*) for your reference.

At the time of writing, the Docker/default profile was deprecated, so we recommend you use runtime/default as the seccomp profile instead.

Table 4-2 shows the options for deploying seccomp profile in Kubernetes via PSP annotations.

Table 4-2. Seccomp options

Value	Description
runtime/default	Default container runtime profile
unconfined	No seccomp profile—this option is default in Kubernetes
localhost/<path>	Your own profile located on node, usually in /var/lib/kubelet/seccomp directory

SELinux

In the recent past, every container runtime breakout (container escape or privilege escalation) was some kind of filesystem breakout (i.e., CVE-2019-5736, CVE-2016-9962, CVE-2015-3627, and more). SELinux mitigates these issues by providing control over who can access the filesystem and the interaction between resources (i.e., user, files, directories, memory, sockets, and more). In the cloud computing context, it makes sense to apply SELinux profiles to workloads to get better isolation and reduce attack surface by limiting filesystem access by the host kernel.

SELinux was originally developed by the National Security Agency in the early 2000s and is predominantly used on Red Hat- and centOS-based distros. The reason SELinux is effective is it provides a MAC, which greatly augments the traditional Linux discretionary access control (DAC) system.

Traditionally with the Linux DAC, users have the ability to change permissions on files, directories, and the process owned by them. And a root user has access to everything. But with SELinux (MAC), each OS resource is assigned a label by the kernel, which is stored as extended file attributes. These labels are used to evaluate SELinux policies inside the kernel to allow any interaction. With SELinux enabled, even a root user in a container won't be able to access a host's files in a mounted volume if the labels are not accurate.

SELinux operates in three modes: Enforcing, Permissive, and Disabled. Enforcing enables SELinux policy enforcement, Permissive provides warnings, and Disabled is to no longer use SELinux policies. The SELinux policies themselves can be further categorized into Targeted and Strict, where Targeted policies apply to particular processes and Strict policies apply to all processes.

The following is the SELinux label for Docker binaries on a host, which consists of `<user:role:type:level>`. Here you will see the type, which is `container_runtime_exec_t`:

```
$ ls -Z /usr/bin/docker*
-rwxr-xr-x. root root system_u:object_r:container_runtime_exec_t:s0
/usr/bin/docker
-rwxr-xr-x. root root system_u:object_r:container_runtime_exec_t:s0
/usr/bin/docker-current
-rwxr-xr-x. root root system_u:object_r:container_runtime_exec_t:s0
/usr/bin/docker-storage-setup
```

To further enhance SELinux, multicategory security (MCS) is used to allow users to label resources with a category. So a file labeled with a category can be accessed by only users or processes of that category.

Once SELinux is enabled, a container runtime like Docker (*https://oreil.ly/WKf97*), podman (*https://oreil.ly/HW4Cc*), or `CRI-O` (*https://oreil.ly/2bOJA*) picks a random MCS label to run the container. These MCS labels consist of two random numbers between 1 and 1023, and they are prefixed with the character "c" (category) and a sensitivity level (i.e., s0). So a complete MCS label looks like "s0:c1,c2." As shown in Figure 4-5, a container won't be able to access a file on a host or Kubernetes volume unless it is labeled correctly as needed. This provides an important isolation between resource interaction, which prevents many security vulnerabilities targeted toward escaping containers.

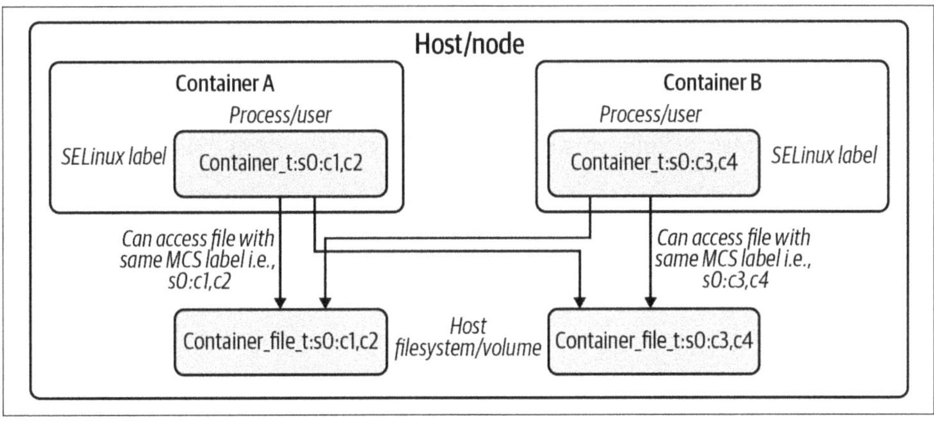

Figure 4-5. SELinux enforcing filesystem access

Next is an example of a pod deployed with SELinux profile; this pod won't be able to access any host volume mount files unless they are labeled so:c123,c456 on host. Even though you see the entire host, the filesystem is mounted on the pod:

```
apiVersion: v1
metadata:
  name: pod-se-linux-label
  namespace: default
  labels:
    app: normal-app
spec:
  containers:
  - name: app-container
    image: alpine:latest
    args: ["sleep", "10000"]
    securityContext:
      seLinuxOptions:
        level: "s0:c123,c456"
  volumes:
    - name: rootfs
      hostPath:
        path: /
```

Table 4-3 lists the CVEs pertaining to container escape that are prevented just by enabling SELinux on hosts. Though SELinux policies can be challenging to maintain, they are critical for a defense-in-depth strategy. Openshift, a Kubernetes distribution, comes with SELinux enabled in its default configuration with targeted policies; for other distros it's worth checking the status.

Table 4-3. CVEs related to container escape

CVE	Description	Blocked by SELinux
CVE-2019-5736	Allows attackers to overwrite host runc binary and consequently obtain host root access	Yes
CVE-2016-9962	RunC exec vulnerability	Yes
CVE-2015-3627	Insecure file-descriptor exploitation	Yes

Kubernetes provides the following options to enforce SELinux in PSPs:

Value	Description
MustRunAs	Need to have seLinuxOptions configured as shown in Figure 4-5.
RunAsAny	No defaults are provided in PSP (can be optionally configured on pod and deployments)

AppArmor

Like SELinux, AppArmor was developed for Debian and Ubuntu operating systems. AppArmor works in a similar way to SELinux, where an AppArmor profile defines what a process has access to. Let's look at an example of an AppArmor profile:

```
#include <tunables/global>
/{usr/,}bin/ping flags=(complain) {
  #include <abstractions/base>
  #include <abstractions/consoles>
  #include <abstractions/nameservice>

  capability net_raw,
  capability setuid,
  network inet raw,

  /bin/ping mixr,
  /etc/modules.conf r,

  # Site-specific additions and overrides. See local/README for details.
  #include <local/bin.ping>
}
```

Here a ping utility has only three capabilities (i.e., net_raw, setuid, and inet raw and read access to /etc/modules.conf). With these permissions a ping utility cannot modify or write to the filesystem (keys, binaries, settings, persistence) or load any modules, which reduces attack surface for the ping utility to perform any malicious activity in case of a compromise.

By default, your Kubernetes runtime like Docker (*https://oreil.ly/WKf97*), podman (*https://oreil.ly/HW4Cc*), or CRI-O (*https://oreil.ly/Cgrep*) provides an AppArmor profile. Docker's runtime profile is provided for your reference (*https://oreil.ly/utKNx*).

Since AppArmor is much more flexible and easy to work with, we recommend having a policy per microservice. Kubernetes provides the following options to enforce these policies via PSP annotations:

Value	Description
runtime/default	Runtime's default policy
localhost/<profile_name>	Apply profile loaded on host, usually in directory /sys/kernel/security/apparmor/profiles
unconfined	No profile will be loaded

Sysctl

`Kubernetes sysctl` (*https://oreil.ly/zwlcG*) allows you to use the sysctl interface to use and configure kernel parameters in your cluster. An example of using sysctls is to manage containers with resource-hungry workloads that need to handle a large number of concurrent connections or need a special parameter set (i.e., IPv6 forwarding) to run efficiently. In such cases, sysctl provides a way to modify kernel behavior only for those workloads without affecting the rest of the cluster.

The sysctls are categorized into two buckets: safe and unsafe. Safe sysctl only affects the containers, but unsafe sysctl affects the container and node it is running on. Sysctl lets administrators set both the sysctl buckets at their discretion.

Let's take an example where a containerized web server needs to handle a high number of concurrent connections and needs to set the net.core.somaxconn value to a higher number than the kernel default. In this case it can be set as follows:

```
apiVersion: v1
kind: Pod
metadata:
  name: sysctl-example
spec:
  securityContext:
    sysctls:
    - name: net.core.somaxconn
      value: "1024"
```

Please note that we recommend that you use node affinity to schedule workloads on nodes that have the sysctl applied, in case you need to use a sysctl that applies to the node. The following example shows how PSPs allow sysctls to be forbidden or allowed:

```
apiVersion: policy/v1beta1
kind: PodSecurityPolicy
metadata:
  name: sysctl-psp
spec:
  allowedUnsafeSysctls:
  - kernel.msg*
  forbiddenSysctls:
  - kernel.shm_rmid_forced
```

Conclusion

In this chapter we covered tools and best practices for defining and implementing your workload runtime security. The most important takeaways are:

- Pod security policies are an excellent way to enable workload controls at workload creation time. They have limitations but can be used effectively.
- You need to pick a solution that is native to Kubernetes for monitoring processes and implement controls based on your threat model for your workloads.
- We recommend you review the various security options that are available in the Linux kernel and leverage the right set of features based on your use case.

CHAPTER 5

Observability

In this chapter we will discuss the difference between monitoring and observability in the context of Kubernetes deployments. We will explain best practices and tools for implementing observability in your Kubernetes cluster. In the next chapter we will cover how you can use observability to secure your cluster.

Observability has been a topic of discussion recently in the Kubernetes community and has garnered a lot of interest. We begin by understanding the difference between monitoring and observability. We then look at why observability is critical to security in a distributed application like Kubernetes, and review tools and reference implementations for observability. While observability is a broad topic and applies to several areas, we will keep the discussion focused on Kubernetes in this chapter. Let's start by looking at monitoring and observability and how they are different.

Monitoring

Monitoring is a known set of measurements in a system that are used to alert for deviations from a normal range. The following are examples of types of data you can monitor in Kubernetes:

- Pod logs
- Network flow logs
- Application flow logs
- Audit logs

Examples of metrics you can monitor include the following:

- Connections per second
- Packets per second, bytes per second
- Application (API) requests per second
- CPU and memory utilization

These logs and metrics can help you identify known failures and provide more information about the symptom to help you remediate the issue.

In order to monitor your Kubernetes cluster, you use techniques like polling and uptime checks depending on the SLAs you need to maintain for their cluster. The following are examples of metrics you could monitor for SLAs:

- Polling of application/API endpoints
- Application response codes (e.g., HTTP or database error codes)
- Application response time (e.g., HTTP duration, database transaction time)
- Node availability for scale-out use cases
- Memory/CPU/disk/IO resources on a node

The other important part of monitoring is alerting. You need an alerting system as part of your monitoring solution that generates alerts for any metric that violates the specified threshold. Tools like Grafana, Prometheus, OpenMetrics, OpenTelemetry, and Fluentd are used as monitoring tools to collect logs and metrics, and generate reports, dashboards, and alerts for Kubernetes clusters. Kubernetes offers several integrations to tools like Opsgenie, PagerDuty, Slack, and JIRA for alert forwarding and management.

Monitoring your production Kubernetes cluster has the following issues:

Amount of log data

In a system like Kubernetes, a node has several pods that run on the host, and each pod comes with its own logs, its own network identity, and its own resources. This means you have logs from the application operation, network flow logs, Kubernetes activity (audit) logs, and application flow logs for each pod. In a non-Kubernetes environment, you typically had an application running on a node and so it would be just one set of logs as opposed to one set of logs per pod running on the node. This multiplies the amount of log data that needs to be collected/inspected. In addition to the per-pod logs, you also need to collect cluster logs from Kubernetes. Typically these are also known as audit logs that provide visibility into Kubernetes cluster activity. The number of logs in the system will make monitoring very resource-intensive and expensive to maintain. Your log collection cluster should not be more expensive to operate than the cluster running your applications!

Monitoring distributed applications
In a Kubernetes cluster, applications are distributed across the Kubernetes cluster network. An application that needs more than one pod (e.g., a deployment set or a service) will have logs for each pod that need to be examined in addition to the context of the set of pods (e.g., scale out, error handling, etc.). We have multiple pods that need to be considered as a group before we generate an alert for the application. Please note the goal is to monitor the application and generate alerts for the application, and generating alerts for pods that are a part of the application independently does not provide an accurate representation of the state of the application. There is also the case of the microservices application, where a single application is deployed as a set of services known as *microservices*, and each microservice is responsible for a part of the functionality of the application. In this case, you need to monitor each microservice as an entity (note a microservice is a set of one or more pods) and then understand which microservices impact any given application transaction. Only then can you report an alert for the application.

Declarative nature of Kubernetes
As we have covered, Kubernetes is declarative and allows you to specify exactly how you want pods to be created and run in the cluster. Kubernetes allows you to specify resource limits for memory, CPU, storage, etc., and you can also create custom resources and specify limits for these resources. The scheduler will find a node that has the required resources and schedule a pod on the node. Kubernetes also monitors usage for pods and will terminate pods that consume more resources than those allocated to them. In addition, Kubernetes provides detailed metrics that can be used to monitor pods and cluster state. For example, you can use a tool like Prometheus (*https://oreil.ly/zzjVG*) that can monitor pods and cluster state and use the metrics, and you can automatically scale pods or other cluster resources with a mechanism known as the Horizontal Pod Autoscaler (*https://oreil.ly/luM5u*). What this means is that Kubernetes as a part of its operation is monitoring and making changes to the cluster to maintain operations as per the configured specification. In this scenario, an alert from monitoring a single metric can be a result of Kubernetes making changes to adapt to the load in the cluster, or it could be a real issue. You need to be able to distinguish between the two scenarios to be able to accurately monitor your application.

Now that we understand monitoring and how it can be implemented and the challenges with using monitoring for a Kubernetes cluster, let's look at observability and how it can help overcome these challenges.

Observability

Observability is defined as the ability to understand the internal state of a system by only looking at external outputs of the system. *Observability Engineering* by Charity Majors et al. (O'Reilly) is an excellent resource to learn more about observability. The book's second chapter discusses monitoring and observability and is very relevant to this discussion.

Observability builds on monitoring and enables you to gain insights about the internal state of your application. For example, in a Kubernetes cluster an unexpected pod restart event may have limited to no impact on services as other instances of the pod may be adequate to handle the load at the time of the restart. A monitoring system will generate an alert that an unexpected pod restart occurred, and an observability system will generate a medium-priority event with the context that an unexpected pod restart occurred but had no impact on the system if there is no other event like application errors at the time of the pod restart. Another example is when an event is generated at the application layer (e.g., duration for HTTP request is larger than the norm). In this scenario, the observability system will provide context for the reason of degradation in application response time (e.g., network layer issue, retransmits, application pod restarts due to resource or other application issues, a Kubernetes infrastructure issue like DNS latency or API server load). As explained previously, an observability system can look at multiple events that impact application state and report application status after considering all of them. Now let's look at how you can implement observability in a Kubernetes system.

How Observability Works for Kubernetes

The declarative nature of Kubernetes helps a lot in implementing an observability system. We recommend that you build a system that is native to Kubernetes and is able to understand operations in a cluster. For example, a system that understands Kubernetes will monitor a pod (e.g., restarts, out of memory, network activity, etc.) but also understand if a pod is a standalone instance or part of a deployment, replica set, or service. It will also know how critical the pod is to the service or deployment (e.g., how the service is configured for scalability and high availability). So when it reports any event related to the pod, it will provide all this context and help you easily make a decision about how you need to respond to the event.

Another thing to remember is that in Kubernetes you can deploy applications as pods that are a part of higher-level constructs like a deployment or a service. In order to appreciate the complexity in implementing observability for these constructs, we will use an example to explain them. When you configure a service, Kubernetes manages all pods associated with the service and ensures that traffic is delivered to available pods that are a part of the service. Let's take a look at an example of service definition from the Kubernetes documentation (*https://oreil.ly/ijVz5*):

```yaml
apiVersion: v1
kind: Service
metadata:
  name: my-service
spec:
  selector:
    app: MyApp
  ports:
    - protocol: TCP
      port: 80
      targetPort: 9376
```

In this example all pods that have the label MyApp and listen on TCP port 9376 become part of the service, and all traffic destined to the service is redirected to these pods. We cover this concept in detail in Chapter 8. So in this scenario, the observability solution should also work to provide insights at the service level. Monitoring a pod in this case is not sufficient. What is needed is that the observability aggregates metrics across all pods in a service and uses the aggregated information for more analytics and alerts.

Now let's look at an example of deployments (*https://oreil.ly/23Eam*) in Kubernetes. Deployments allow you to manage pods and replica sets (replicas of a pod, typically used for scaling and high availability). The following is an example configuration for a deployment in Kubernetes:

```yaml
apiVersion: apps/v1
kind: Deployment
metadata:
  name: nginx-deployment
  labels:
    app: nginx
spec:
  replicas: 3
  selector:
    matchLabels:
      app: nginx
  template:
    metadata:
      labels:
        app: nginx
    spec:
      containers:
      - name: nginx
        image: nginx:1.14.2
        ports:
        - containerPort: 80
```

This configuration will create a deployment for nginx with three replica pods with the configured metadata and specification. Kubernetes has a deployment controller to ensure that all pods and replicas that are a part of the deployment are available. There are several other benefits, like rolling updates, autoscaling, etc., that can be achieved by using the deployment resource in Kubernetes. In such a scenario for observability, the tool you use should look at the activity of all pods (replicas) in a deployment as an aggregate (e.g., all traffic to/from pods in a deployment, pod restarts and their effect on a deployment, etc.). Monitoring and alerting for each pod will not be sufficient to understand how the deployment is operating.

In both these examples it is clear that the collection of metrics needs to be in the context of Kubernetes. Instead of collecting all data and metrics at a pod-level granularity, the collection engine should collect data at a deployment- or service-level granularity when applicable to deliver an accurate representation of the state of the deployment or service. Remember, Kubernetes abstracts pod-level details, and so we need to focus on measuring and alerting at a higher level than pods. Aggregation of data at a deployment and service level will reduce the number of logs you need to collect all the time and address the concern of the costs associated with a large number of logs. Please note the tool needs to have the ability to drill down and capture pod-level details when the operator needs to analyze an issue. We will cover this later in this chapter when we discuss data collection.

Now that we understand how we can leverage the declarative nature of Kubernetes to help with observability and reduce the amount of log data we need to collect and generate relevant alerts, let's explore the distributed nature of Kubernetes and its impact on observability.

In a microservices-based application deployment, a single application comprises several microservices that are deployed in a Kubernetes cluster. This means that in order to service a single transaction from the user, one or more services need to interact with each other, resulting in one or more subtransactions. A great example of a sample microservices application is the Google online boutique demo microservices application (*https://oreil.ly/wx7bj*). Figure 5-1 shows the architecture for this application.

Figure 5-1 shows how an online boutique application can be deployed as microservices in Kubernetes. There are 11 microservices, each responsible for some aspect of the application. We encourage you to review this application as we will use it to demonstrate how you can implement observability later in the chapter. If you look at the checkout transaction, a user makes a request to the frontend service, which then makes a request to the checkout service. The checkout service needs to interact with several services (e.g., PaymentService, Shipping Service, CurrencyService, EmailService, ProductCatalog Service, CartService) to complete the transaction. So in this scenario if we see our HTTP application log indicate a larger-than-expected duration for the checkout process API response time, we will need to review each of the subtransactions and see if there is an issue with each one and what the issue is (an application issue, network issue, etc.). Another thing that makes this complicated is the fact that each subtransaction is asynchronous and each microservice is serving several independent transactions simultaneously. In such a scenario you need to use a technique known as *distributed tracing* to trace the flow of a single transaction across a set of microservices. Distributed tracing can happen by instrumenting the application or instrumenting the kernel. We will cover distributed tracing later in the chapter.

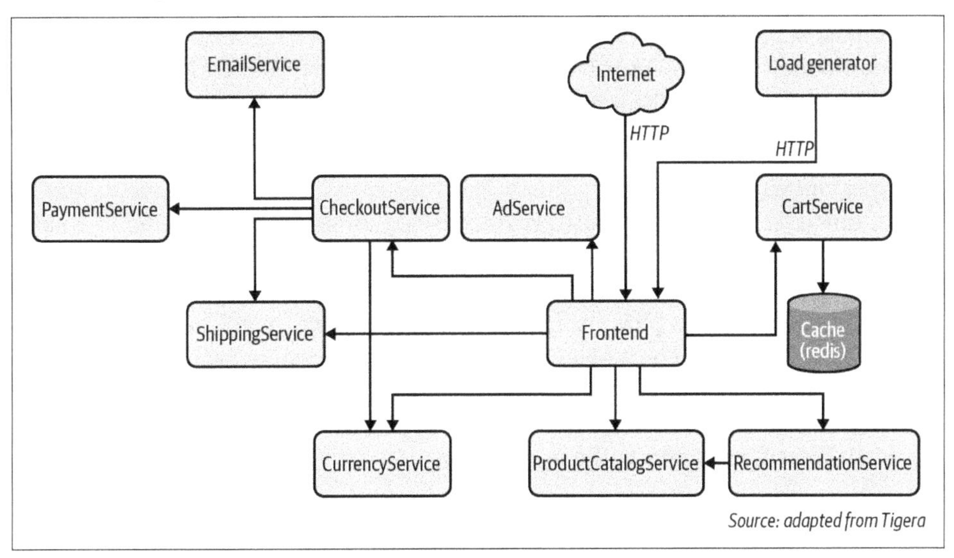

Figure 5-1. Architecture of the Google microservices demo application

Now that we understand observability and how you should think about it for a Kubernetes cluster, let's look at the components for an observability tool for Kubernetes. Figure 5-2 shows a block diagram of the various components of an observability tool for Kuebrnetes.

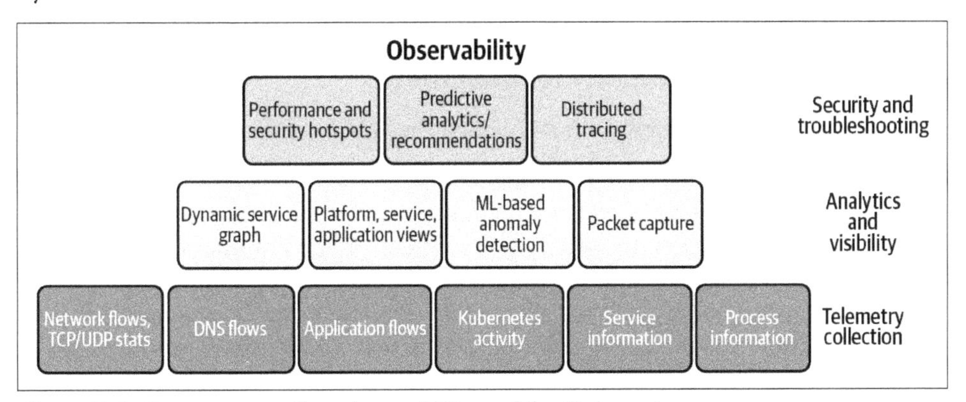

Figure 5-2. Components of an observability tool for Kubernetes

Figure 5-2 shows that you need the following components for your observability implementation:

Telemetry collection
: As mentioned, your observability solution needs to collect telemetry data from various sensors in your cluster. It needs to be distributed and Kubernetes-native. It must support sensors across all layers, from L3 to L7. It also needs to collect information about Kubernetes infrastructure (e.g., DNS and API server logs) and Kubernetes activity (these are known as audit logs). As described, this information must be collected in the context of deployments and services.

Analytics and visibility
: In this layer, the system must provide visualizations that are specific to Kubernetes operations (e.g., service graph, Kubernetes platform view, application views). We will cover some common visualizations that are native to Kubernetes. We recommend you pick a solution that leverages machine learning techniques for baselining and reporting anomalies. Finally, the system needs to support the ability for operators to enable pod-to-pod packet capture (note that this is not the same as enabling packet capture on the host interface, as the pod-level visibility is lost). We will cover this in the next section.

Security and troubleshooting applications
: The observability system you implement must support distributed tracing as described in the previous section to help troubleshoot applications. We also recommend the use of advanced machine learning techniques to understand Kubernetes cluster behavior and predict performance or security concerns. Please note that this is a new area and there is ongoing innovation in it.

Now that we have covered what is needed to implement observability in a Kubernetes cluster, let's review each of the components in detail.

Implementing Observability for Kubernetes

In this section we will review each component needed to build an effective observability system in Kubernetes.

You should think of log collection as a set of sensors that are distributed in your cluster. You need to ensure that the sensors are efficient and do not interfere with system operation (e.g., adding latency). We will cover methods of collection later in this section that will show how you can efficiently collect metrics. You should consider deploying sensors (or collecting information) across all the layers of the stack, as shown in Figure 5-2. Kubernetes audit logs are an excellent source of information to understand the complete life cycle of various Kubernetes resources. In addition to audit logs, Kubernetes provides a variety of options for monitoring (*https://oreil.ly/FSwUs*). Next you need to focus on traffic flow logs (Layer 3/Layer 4) to understand the operation of the Kubernetes cluster network. Given the declarative nature of Kubernetes, it is important to collect logs related to application flows (e.g., HTTP or MySQL), where the logs provide visibility into the application behavior (e.g., response time, availability, etc.) as seen by the user. In order to help with troubleshooting, you should also collect logs related to Kubernetes cluster infrastructure (e.g., API server, DNS). Some advanced troubleshooting systems also collect information from the Linux kernel that is a result of activity by a pod (e.g., process information, socket stats for a flow initiated by a pod) and provide a way to enable packet capture (raw packets) for pod-to-pod traffic. The following describes what you should collect for each:

Kubernetes audit logs
: Kubernetes provides the ability to collect and monitor activity. Here is an excellent guide (*https://oreil.ly/aKmCE*) that explains how you can control what to collect and also mechanisms for logging and alerting. We suggest you review what you need to collect and set the audit policy carefully—we recommend against just collecting everything. For example, you should log API requests, usernames, RBAC, decisions, request verbs, the client (user-agent) that made the request, and response codes for API requests. We will show a sample Kubernetes activity dashboard in the visualization section.

Network flow logs
: Network flow logs (Layer 3/Layer 4) are key to understanding the Kubernetes cluster network operation. Typically these include the five-tuple (source and destination IP addresses/ports and port). It is also important to collect Kubernetes metadata associated with pods (source and destination namespaces, pod names, labels associated with pods, host on which the pods were running) and aggregate bytes/packets for each flow. Note this can result in a large amount of flow data, as there could be a large number of pods on a node. We will address this in the following section about aggregation at collection time.

DNS flow logs
: Along with the API server, the DNS server is a critical part of the Kubernetes cluster and is used by applications to resolve domain names in order to connect other services/pods as a part of normal operation. An issue with the DNS server can impact several applications in the cluster. It is important to collect information from the client's perspective. You should log DNS requests by pods that capture request count, latency, which DNS server was used to resolve the request, the DNS response code, and the response. This should be collected with Kubernetes metadata (e.g., namespace, pod name, labels, etc.), as this will help associate the DNS issue with a service and facilitate further troubleshooting.

Application logs
: As explained, the collection of application logs (HTTP, MySQL) is very important in a declarative system like Kubernetes. These logs provide a view into the

user experience (e.g., response time or availability). The logs will be application-specific information but must include response codes (status), response time, and other application-specific context. For example, for HTTP requests, you should log domains (part of the URL), user agent, number of requests, HTTP response codes, and in some cases complete URL paths. Again, logs should include Kubernetes metadata (e.g., namespace, service, labels, pod names, etc.).

Process information and socket stats
: As mentioned, these stats are not part of typical observability implementations, but we recommend that you consider collecting these stats as they provide a more comprehensive view of the Kubernetes cluster operation. For example, if you can get information about processes (that run in a pod), this can be an excellent way to correlate with application performance data (e.g., co-relating memory usage, or garbage collection events in a Java-based application to response time and network activity initiated by the process). Socket stats are details of a TCP flow between two endpoints (e.g., network round-trip time, TCP congestion windows, TCP retransmits, etc.). These stats when associated with pods can provide a view into the impact of the underlying network on pod-to-pod communication.

Now that we have covered what you need to collect for a complete observability solution, let's look at the tools and techniques available to implement collection. Figure 5-3 is an example reference implementation to show how you can implement collection on a node in your Kubernetes cluster.

Figure 5-3 shows a node in your Kubernetes cluster that has applications deployed as services, deployments, and pods in namespaces as you would see in a typical Kubernetes cluster. In order to facilitate collection, a few components are added as shown in the observability components section, and it shows a few additions to the Linux kernel to facilitate collection. Let's explore the functions of each of these components.

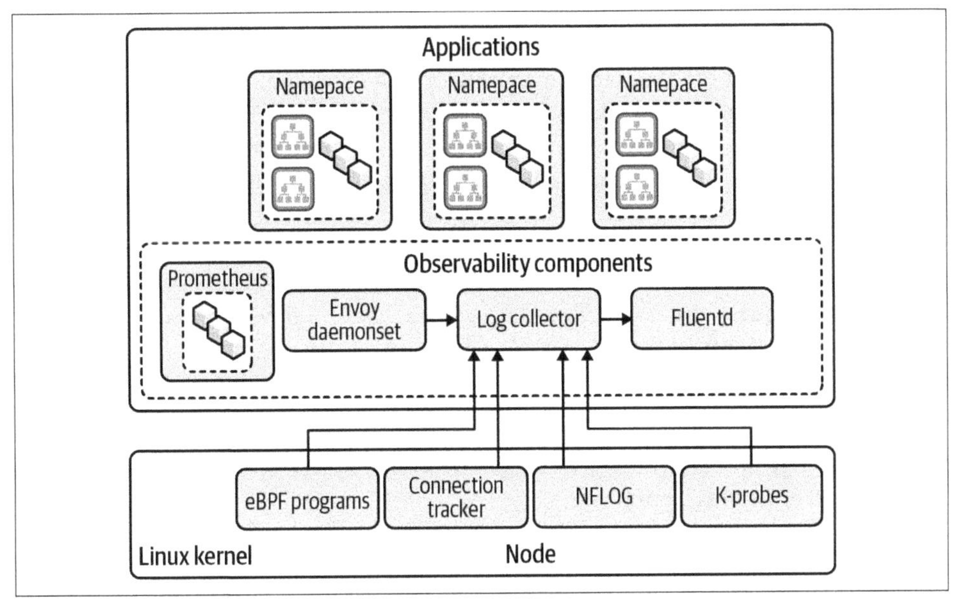

Figure 5-3. Reference implementation for collection on a node

Linux Kernel Tools

The Linux kernel offers several options that you can use to help with data collection. It is very important that the tool you use leverages these tools instead of focusing on processing raw logs that are generated by other tools:

eBPF programs and kprobes
: eBPF stands for extended Berkley Packet Filter. It is an exciting technology that can be used for collection and observability. It was originally designed for packet filtering, but was then extended to allow adding programs to various hooks in the kernel to be used as trace points. In case you are using an eBPF-based dataplane, the eBPF programs that are managing the packet path will also provide packet and flow information. We recommend reading Brendan Gregg's blog post "Linux Extended BPF (eBPF) Tracing Tools" (*https://oreil.ly/s54kG*) to understand how to use eBPF for performance and tracing. In the context of this discussion, you can attach an eBPF program to a kernel probe (kprobe), which is essentially a trace point that is triggered and executes the program whenever the code executes the function for which the kprobe is registered. The kernel documentation for kprobes (*https://oreil.ly/hLziH*) provides more details. This is a great way to get information from the Linux kernel for observability.

NFLOG and conntrack

If you are using the standard Linux networking dataplane (iptables-based), there are tools available to track packets and network flows. We recommend using NFLOG, which is a mechanism to be used in conjunction with iptables to log packets. You can review the details in the iptables documentation; at a high level NFLOG can be set as a target for an iptables rule, and it will log packets via a netlink socket on a multicast address that a user space process can subscribe to and collect packets from. Conntrack is another module used in conjunction with iptables to query the connection state of a packet or a flow, and it can be used to update statistics for a flow.

We recommend you review options (e.g., Net Filter) that the Linux kernel provides and leverage them in sensors that are used to collect information. This is very important as it will be an efficient way to collect data, since these options provided by the Linux kernel are highly optimized.

Observability Components

Now that we understand how to collect data from the Linux kernel, let's look at how this data needs to be processed in user space to ensure we have an effective observability solution:

Log collector

This is a very important component in the system. The goal of this component is to add context from the Kubernetes cluster to the data collected from other sensors—for example, to add pod metadata (name, namespace, label, etc.) to source and destination IP addresses, respectively, from a network flow. This is how you can add Kubernetes context to raw network flow logs. Likewise, any data you collect from kernel probes can also be enriched by adding relevant Kubernetes metadata. This way you can have log data that associates activity in the kernel to objects in your Kubernetes cluster (e.g., pods, services, deployments). It is critical for you to be able derive insights about your Kubernetes cluster operation. Please note that this component is something you need to implement, or you must ensure that the tool you choose for observability has this functionality. It is a critical part of your observability implementation.

Envoy (proxy)

We discussed the importance of having a collection of application-specific data, and for this we recommend that you use Envoy (*https://oreil.ly/0niF8*), a well-known proxy that is used to analyze application protocols and log application transaction flows (e.g., HTTP transactions on a single HTTP connection). Please note that Envoy can be used as a sidecar pattern where it attaches to every pod as a sidecar and tracks packets to/from the pod. It can also be deployed as a daemonset (a transparent proxy) where you can use the dataplane to redirect traffic to pass through an envoy instance running on the host. We strongly recommend using Envoy with this latter configuration, as using the sidecar pattern has security concerns and can be disruptive to applications. In the context of this discussion, the Envoy daemonset will be the source of application flow logs to the log collector. The log collector can now use the pod metadata (name, namespace, labels, deployments, services, IP addresses) to correlate this data with the data received from the kernel and further enrich it with application data.

Fluentd

Note that the data collection discussed so far is processed by the log collector on every node in the cluster. You need to implement a mechanism to send the data from all nodes to a datastore or security information and event management (SIEM), where it can be picked up by analytics and visualization tools. Fluentd (*https://oreil.ly/11s22*) is an excellent option to send collected data to the datastore of your choice. It offers excellent integrations and is a tool that is Kubernetes native. There are other options available, but we recommend you use Fluentd for shipping collected log data to a data store.

Prometheus

We've discussed how you collect flow logs; now we need a component for the collection of metrics and alerting. Prometheus, a Kubernetes-native tool, is a great choice for metrics collection and alerting. It's deployed as endpoints that scrape metrics and send them to a time-series database that's a part of the Prometheus server for analysis and query. You can also define alerts for data sent to the Prometheus server. It's a widely used option and has integrations to dashboards and alerting tools. We recommend you consider it as an option for your cluster.

We hope that this discussion has given you an idea of how you can implement data collection for your Kubernetes cluster. Now let's look at aggregation and correlation.

Aggregation and Correlation

In the previous section we covered data collection and discussed how you can collect data from various sources in your cluster (API server, network flows, kernel probes, application flows). This is very useful, but we still need to address the concern of data volume if we keep the collection at pod-level granularity. Another thing to note is that the data volume concern multiples if we keep data from various sources separate and then associate it at query time. You can say that it's better to keep as much raw data as possible, and there are efficient tools to query and aggregate data after collection (offline), so why not use that approach? Yes, that is a valid point, but there are a couple of things to think about. The large volume of data would mean aggregation and query time joins of data will be resource-intensive (it can very well be more expensive to operate your data collection system than your Kubernetes cluster!). Also, given the ephemeral nature of Kubernetes (pod life cycles can be very short), the latency in analyzing data offline prevents any kind of reasonable response to data collected to mitigate the issue reported by the data. In some cases, if correlation is not done at collection time, it will not be possible to associate two different collections. For example, you cannot collect a list of policies and a list of flows and then associate the policy with a flow offline without rerunning the policy evaluation.

We also discussed the declarative nature of Kubernetes and how a deployment and a service are higher-level constructs than a pod. In such a scenario, we recommend that you consider the aggregation of data at a deployment or a service level. This means data from all pods for a service is aggregated; you would collect data between deployments and data to services as a default option. This will give you the right level of granularity. You can provide an option to reduce the aggregation to collect pod-level data as a drill-down action or in response to an event. This way you can address the concern about large amounts of log data collected and the associated processing cost. Also, the data collection is more Kubernetes-native as Kubernetes monitors deployments/services as a unit and makes adjustments to ensure the deployment/service is operating as per the specification.

In the data collection section we discussed the log collector component that receives data from various sources. It could be used as a source to correlate data at collection time, so you don't have to do any additional correlation after data collection, and you also benefit from not having to collect redundant data for each source. For example, if the kprobe in the kernel collects socket data for five-tuple (IP addresses, ports, protocol), and the NFLOG provides other information like bytes and packets for the same five-tuple, the log collector can create a single log with the five-tuple, the Kubernetes metadata, the network flow data, and the socket statistics. This will provide logs with very high context and low occupancy for collection and processing.

Now let's go back to the Google online boutique example and see a sample of what a log will look like with aggregation and correlation of kernel and network flow data. The sample log is generated using the collection and aggregation concepts described previously for a transaction between the frontend service and the currencyservice of the application. It is a gRPC-based transaction:


```
{
    "_id": "YTBT5HkBf0waR4u9Z0U3",
    "_score": 3,
    "_type": "_doc",
    "start_time": 1623033303,
    "end_time": 1623033334,
    "source_ip": "10.57.209.32",
    "source_name": "frontend-6f794fbff7-58qrq",
    "source_name_aggr": "frontend-6f794fbff7-*",
    "source_namespace": "onlinebotique",
    "source_port": null,
    "source_type": "wep",
```

```
    "source_labels": [
        "app=frontend",
        "pod-template-hash=6f794fbff7"
    ],
    "dest_ip": "10.57.209.29",
    "dest_name": "currencyservice-7fd6c64-t2zvl",
    "dest_name_aggr": "currencyservice-7fd6c64-*",
    "dest_namespace": "onlinebotique",
    "dest_service_namespace": "onlinebotique",
    "dest_service_name": "currencyservice",
    "dest_service_port": "grpc",
    "dest_port": 7000,
    "dest_type": "wep",
    "dest_labels": [
        "app=currencyservice",
        "pod-template-hash=7fd6c64"
    ],
    "proto": "tcp",
    "action": "allow",
    "reporter": "src",
    "policies": [
        "1|platform|platform.allow-kube-dns|pass",
        "2|__PROFILE__|__PROFILE__.kns.hipstershop|allow",
        "0|security|security.pass|pass"
    ],
    "bytes_in": 68437,
    "bytes_out": 81760,
    "num_flows": 1,
    "num_flows_started": 0,
    "num_flows_completed": 0,
    "packets_in": 656,
    "packets_out": 861,
    "http_requests_allowed_in": 0,
    "http_requests_denied_in": 0,
    "process_name": "wrk:worker_0",
    "num_process_names": 1,
    "process_id": "26446",
    "num_process_ids": 1,
    "tcp_mean_send_congestion_window": 10,
    "tcp_min_send_congestion_window": 10,
    "tcp_mean_smooth_rtt": 9303,
    "tcp_max_smooth_rtt": 13537,
    "tcp_mean_min_rtt": 107,
    "tcp_max_min_rtt": 107,
    "tcp_mean_mss": 1408,
    "tcp_min_mss": 1408,
    "tcp_total_retransmissions": 0,
    "tcp_lost_packets": 0,
    "tcp_unrecovered_to": 0,
    "host": "gke-v2y0ly8k-logging-default-pool-e0c7499d-76z8",
    "@timestamp": 1623033334000
}
```

This is an example of a flow log from Calico Enterprise. There are a few things to note about the log: It aggregates data from all pods backing the frontend service (`frontend-6f794fbff7-*`) and all pods belonging to the currencyservice (`currencyservice-7fd6c64-*`). The data from the kprobe and socket statistics are aggregated as mean, min, and max for each metric for the data between services. The process ID and the process name received from the kernel are correlated with the other data, and we also see the network policy action and the network policies impacting the flow correlated with other data. This is an example of what you want to achieve for data collection in your Kubernetes cluster!

Now that we have covered how to collect, aggregate, and correlate data in a Kubernetes-native manner, let's explore visualization of data.

Visualization

There are some great tools that support the visualization of the data collected. For example, Prometheus offers an integration with Grafana that provides very good dashboards to visualize data. There are also some commercial tools like Datadog, New Relic, and Calico Enterprise that support the collection and visualization of data. We will cover a few common visualizations that are useful for Kubernetes clusters.

Service Graph

This is a representation of your Kubernetes cluster as a graph showing services in a Kubernetes cluster and interactions between them. If we go back to the Google microservices online boutique example, Figure 5-4 shows the online boutique application implemented and represented as a service graph.

Figure 5-4 is a visualization of the online boutique namespace as a service graph, with the nodes representing services and pods backing a service or a group of pods either standalone or as a part of a deployment. The edges show network activity and policy action. The graph is interactive and allows you to pick a service (e.g., frontend service) and allows the viewing of detailed logs collected for the service. Figure 5-5 shows a summarized view of all collected data for the service selected (frontend).

Figure 5-5 shows a detailed view of the frontend service as a drill-down—it shows information from all sources in one view, so it's very easy to analyze the operation of the service.

Service graph is a very common pattern to represent Kubernetes cluster topology. There are several tools that provide this view, such as Kiali, Datadog, and Calico Enterprise.

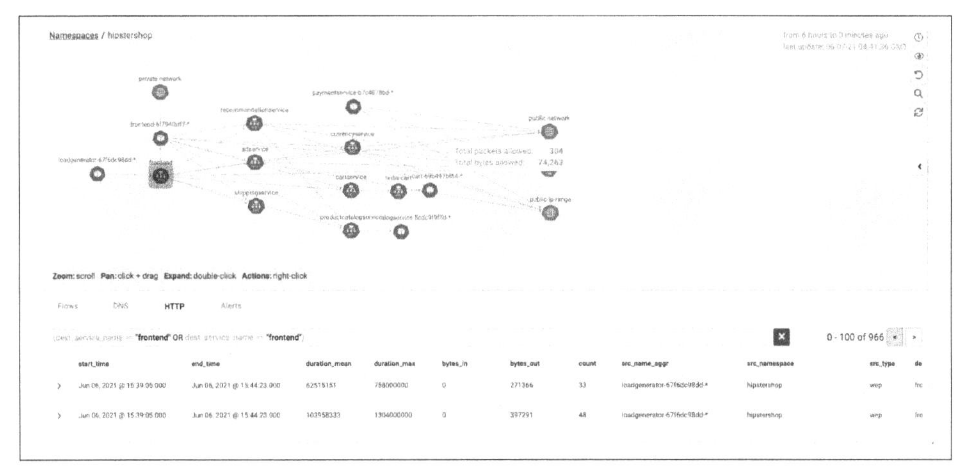

Figure 5-4. Service graph representation of the online boutique application

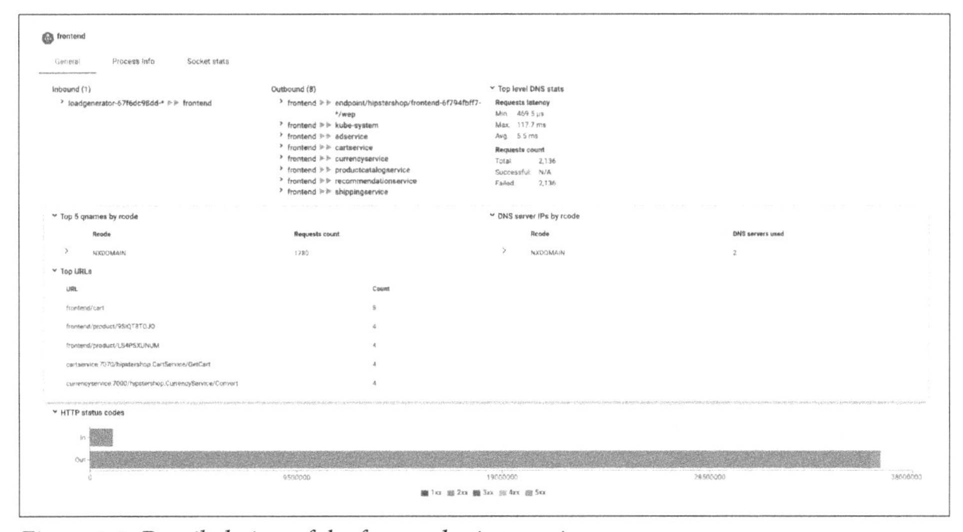

Figure 5-5. Detailed view of the frontend microservice

Visualization of Network Flows

Figure 5-6 shows a common pattern used to visualize flows. This is ring-based visualization, where each ring represents an aggregation level. In the example shown in Figure 5-5, the outermost ring represents a namespace and all flows within the namespaces. Selecting a ring in the middle shows all flows for a service, and selecting the innermost ring shows all flows for pods backing the service. The panel on the right is a selector to enable more-granular views using filtering and details like flows and policy action for the selection. This is an excellent way to visualize network flows in your cluster.

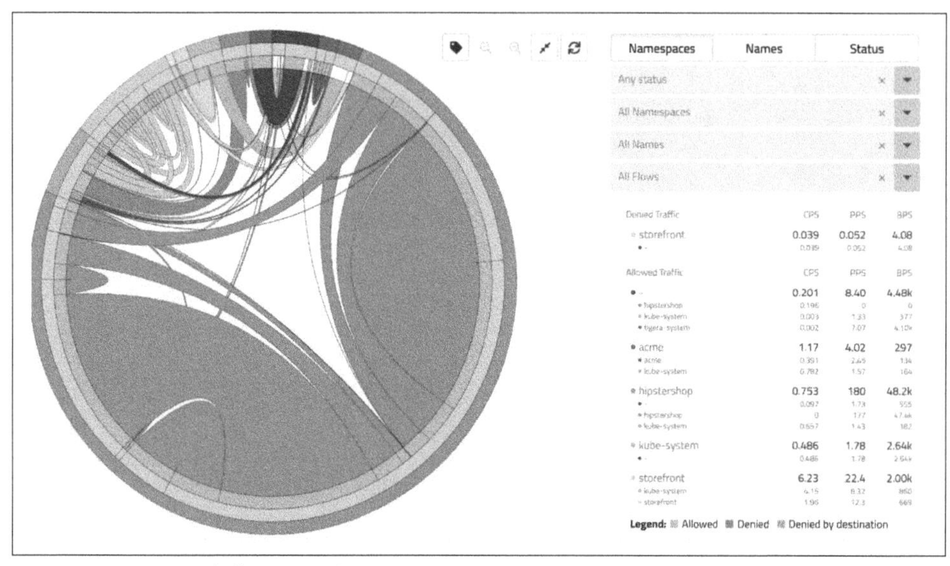

Figure 5-6. Network flow visualization

In this section we have covered some common visualization patterns and tried to show how they can be applied to Kubernetes. Please note that there are several visualizations that can be applied to Kubernetes; these are examples to show how you can represent data collected in a Kubernetes cluster.

Now that we have covered data collection, aggregation, correlation, and visualization, let's explore some advanced topics to utilize the data collected to derive insights into the operation of the Kubernetes cluster.

Analytics and Troubleshooting

In this section we will explore analytics applications that leverage the collection, aggregation, and correlation components to help provide additional insights. Note that there are many applications that can be built to leverage the context-rich data in a Kubernetes cluster. We cover some applications as examples.

Distributed Tracing

We explained distributed tracing before and discussed its importance in a microservices-based architecture, where it is critical to trace a single user request across multiple transactions that need to happen between various microservices. There are two well-known approaches to implementing distributed tracing,

Instrument transaction request headers
In this method the HTTP headers are instrumented with a request ID, and the request ID is preserved in headers across calls to various other services. Envoy (*https://oreil.ly/jprHR*) is a very popular tool used to implement distributed tracing. It supports integrations with other well-known application tracers like Lightstep and AWS X-Ray. We recommend that you use Envoy if you are fine with instrumenting applications to add and preserve the request ID across calls between microservices.

eBPF and kprobes
In the method described for using Envoy, there is a change required to the application traffic. It is possible to implement distributed tracing for service-to-service calls using eBPF and Linux kernel probes. You can attach eBPF programs to kprobes/uprobes and other trace points in the kernel and build a distributed tracing application. Note the detailed implementation of such an application is beyond the scope of this book, but we wanted to mention this as an option for distributed tracing in case you are wary of altering application traffic.

Now that we have covered distributed tracing, let's look at how you can implement packet capture in your Kubernetes cluster.

Packet Capture

In your Kubernetes cluster we recommend that you implement or pick a tool that supports raw packet captures between pods. The tool should support a selector-based packet capture (e.g., pod labels) and role-based access control to enable and view packet captures. This is a simple yet very effective feature that can be used as a response action to an event (e.g., increased application latency) to analyze raw packet flows to understand the issue and find the root cause. In order to implement raw packet captures, we recommend using libpcap (*https://oreil.ly/c2UFJ*), which supports the ability to capture packets on an interface on Linux systems.

Conclusion

In this chapter we covered what observability is and how to implement it for your Kubernetes cluster. The following are the highlights of this chapter:

- Monitoring needs to be a part of your observability strategy; monitoring alone is not sufficient.
- It is important to leverage the declarative nature of Kubernetes when you implement an observability solution.

- The key components for implementing observability for your Kubernetes cluster are log collection, log aggregation and correlation, visualization, distributed tracing, and analytics.
- You must implement your observability using a tool that is native to Kubernetes.
- You should use tools available in the Linux kernel to drive efficient collection and aggregation of data (e.g., NFLOG, eBPF-based probes).

CHAPTER 6

Observability and Security

This chapter will explain how an observability platform can help improve the security of your Kubernetes cluster. We will cover the following topics:

Alerting
: In Chapter 5 we covered best practices for implementing log collection. In this chapter, we will focus on how to build a system that helps generate high-fidelity alerts. We will also discuss the use of machine learning for anomaly detection.

Security operations center
: We will review a reference implementation of a security operations center (SOC) and how observability can help you build an SOC for your Kubernetes cluster.

Behavioral analytics
: We will cover the concept of user and entity behavior analytics (UEBA) and how to implement it in your Kubernetes cluster.

Alerting

In the previous chapter we discussed how to implement logging for your Kubernetes cluster. An effective alerting system must include the following:

- The system should be able to automatically run queries across various log data sources (e.g., Kubernetes activity logs, network logs, application logs, DNS logs, etc.).
- The system must be able to support a state machine that is used to generate events for a specified number of threshold violations in a specified duration. The system must also support setting a time period for the query (known as *look-back*). We will cover an example in the next section.

- The system must be able to export actionable alerts to external security, information, and event management (SIEM) so that they can be a part of the incident response process in an enterprise.

Alerting systems are available from major cloud providers that can help you define alerts on logs collected in the cloud provider environment. Google Cloud has an excellent resource to learn about its alerting capabilities (*https://oreil.ly/DyLh3*). Amazon Web Services (AWS) also has similar alerting capabilities. These alerts work on logs collected in the cloud provider's logging system and allow you to define rules to trigger alerts based on thresholds. For example, the number of API calls to an API endpoint in a given time period can be an indicator of a potential denial of service (DoS) attack. While these systems are good for general logging and alerting, a log collection system that is native to Kubernetes, like the one we covered in Chapter 5, is necessary to detect security-based events in your Kubernetes cluster, as it correlates data at the time of collection and makes it easy to define alerts on one log source. Also, an alerting system native to Kubernetes will help you define alerts using Kubernetes constructs like deployments, labels, etc., as it will be able to enhance log data with the right context so queries are simple. (For example, you do not need to join a set of labels to services and labels to IPs in network flow logs to query network activity for a service.)

Figures 6-1 and 6-2 show you an example of how you can define an alert to detect lateral movement in your Kubernetes cluster.

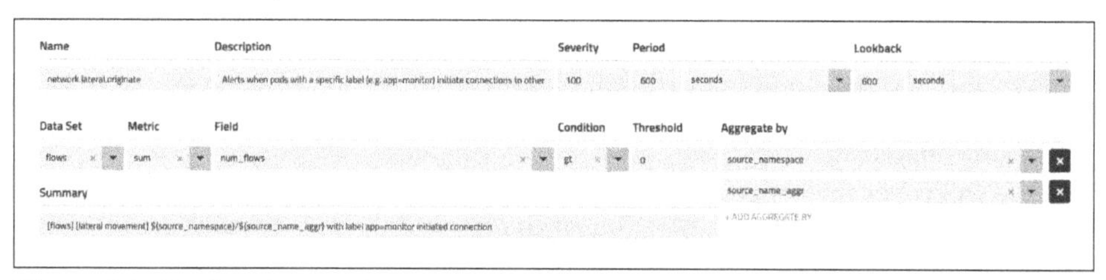

Figure 6-1. Configuring an alert operation

Figure 6-1 shows how you can configure the operation of an alert and metadata like name, description, severity, and time period to poll for the data and the lookback period, which is how far back the system will look when querying the data. You can also define thresholds for occurrences of threshold violations before an alert is triggered. It is also important to be able to configure the format of the output; in this example the data will be aggregated by the source of the traffic (namespace and deployment) when the alert is reported. The output alert data helps facilitate the management of the alert by downstream systems (e.g., SIEM).

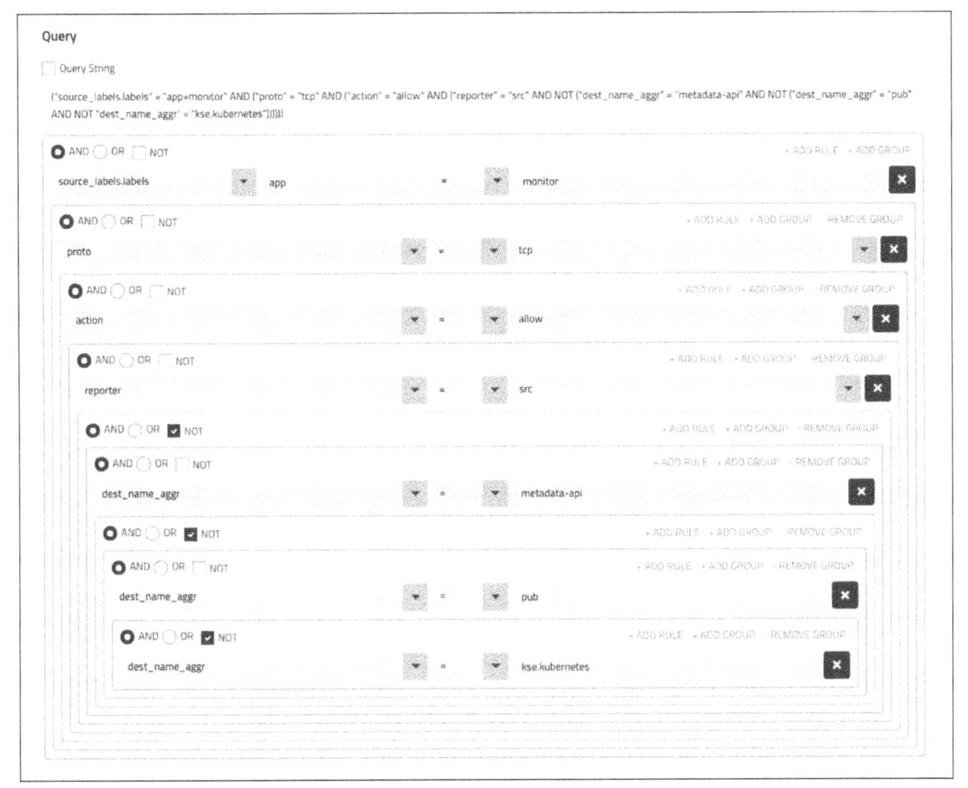

Figure 6-2. Query configuration for an alert

Figure 6-2 shows an example of query configuration for the alert. The things to note in the query are the ability to use Kubernetes metadata (e.g., labels) with network flow activity (e.g., destination, protocol) and policy verdict (e.g., action) in a single query. This gives you a lot of flexibility in defining alerts that are effective in your Kubernetes cluster. Please note that this is a representative example of how you should think about building an effective alerting system. In addition to the cloud provider's alerting systems, there are several other tools like Datadog, Sysdig, and Calico Enterprise that provide Kubernetes-native alerting systems.

The alerting systems we covered previously are great at detecting and reporting alerts when your system has predictable behavior and you can easily define thresholds for normal activity of the system. It would be great for an alerting system to be able to "learn" the behavior of the system and be able to dynamically define thresholds; this will help generate high-fidelity alerts and reduce false positives due to thresholds not changing with the state of the system. Let's explore how machine learning can be used to help with this issue.

Machine Learning

Machine learning fundamentals and how it works are outside the scope of this book. In this section we will review a few concepts of machine learning that will show us how it can help with learning the behavior of a given metric and alerting on deviations from the expected behavior. Before we do that, let's review the high-level techniques in machine learning:

Supervised learning
: This is a technique where the system is trained by labeling test data over a period of time. It allows the system to use the learnings to classify new data and predict outcomes.

Unsupervised learning
: This is a technique where algorithms are used to detect and classify patterns in data that is not labeled. Please note that there are many resources to understand these concepts; one example is Julianna Delua's"Supervised vs. Unsupervised Learning: What's the Difference?" (*https://oreil.ly/9aBfa*). Given the ephemeral nature of entities (e.g., pods) in a Kubernetes cluster and our goal of detecting anomalies in the data generated from that activity, we recommend using the technique of unsupervised learning for detecting anomalies.

Baselining
: Baselining is a technique used to build a model in machine learning that can continuously predict values for a given metric (e.g., connections per second) and detect anomalies (deviations) from the expected value. CMU ML's blog post "3–Baselines" (*https://oreil.ly/zpeNO*) is a great resource for understanding how baselines work and the different types of models that can be built using baselining. As mentioned in the blog, it is possible to create simple models that are very effective in achieving human-level performance. This is exactly what we want in our alerting system: The system should automatically define thresholds and alert on deviations from the baseline.

Now that we understand the high-level techniques we should use, let's look at some example machine learning jobs that help with implementing observability and securing a Kubernetes cluster. In a dynamic environment like Kubernetes, where workloads are ephemeral and can be restarted/scheduled on a different node, it is not practical in most cases to use a rule-based engine to detect anomalies. What is needed is the anomaly-detection engine layered over a machine learning engine that reports deviations from the baseline for any given metric.

Examples of Machine Learning Jobs

How to create a machine learning model is outside the scope of this book; you should have the data science team build models for your deployments. Major cloud providers like Google Cloud offer a service to build machine learning models for Kubernetes workloads (*https://oreil.ly/0JxKq*) that can help the data science team implement the right ML model. The following are some examples of ML jobs that are effective to detect anomalous events in your cluster:

IP sweep detection
: The job looks for pods in your cluster that are sending packets to many destinations. This may indicate an attacker has gained control of a pod and is gathering reconnaissance on what else they can reach. The job compares pods both with other pods in their replica set and with other pods in the cluster generally.

Port scan detection
: The job looks for pods in your cluster that are sending packets to one destination on multiple ports. This may indicate an attacker has gained control of a pod and is gathering reconnaissance on what else they can reach. The job compares pods both with other pods in their replica set and with other pods in the cluster generally.

Service bytes anomaly
: The job looks for services that receive/send an anomalously high amount of data. This could indicate a denial of service attack, data exfiltrating, or other attacks. The job looks for services that are unusual with respect to their replica set, and replica sets that are unusual with respect to the rest of the cluster.

Process restarts anomaly
: The job looks for pods with an excessive number of the process restarts. This could indicate problems with the processes, such as resource problems or attacks. The job looks for pods that are unusual with respect to their process restart behavior.

DNS latency anomaly
: The job looks for the clients that have too-high latency of DNS requests. This could indicate a denial of service attack.

L7 latency anomaly
: The job looks for the pods that have too-high latency of L7 requests. All HTTP requests are measured here. This anomaly could indicate a denial of service attack or other attacks.

HTTP connection spike anomaly
: The job looks for services that get too many HTTP inbound connections. This anomaly could indicate a denial of service attack.

This list gives you examples of jobs you can use to detect anomalies. You can use this resource offered by Google Cloud (*https://oreil.ly/1FaTm*) to build a machine learning job for your Kubernetes cluster. Note the descriptions of each job rely on a set of context-rich logs that are native to Kubernetes (e.g., comparing pods to other pods in a replica set, using bytes sent to/from a service).

Now that we have covered how to build an effective alerting system to detect and report anomalies, let's look at an example implementation of a security operations center for your Kubernetes cluster.

Security Operations Center

In this section we will review a reference implementation for a security operations center (SOC) for a SaaS service based on Kubernetes. An SOC is used to detect and respond to security events; we will explore how to leverage observability when you implement an SOC for your Kubernetes-based services. Note this is a sample and should be used as an example to guide your implementation. When you implement this in production, you should use these concepts but will need to design and implement an SOC suited to your use case. Figure 6-3 shows an SOC implementation for your service hosted in Google Cloud.

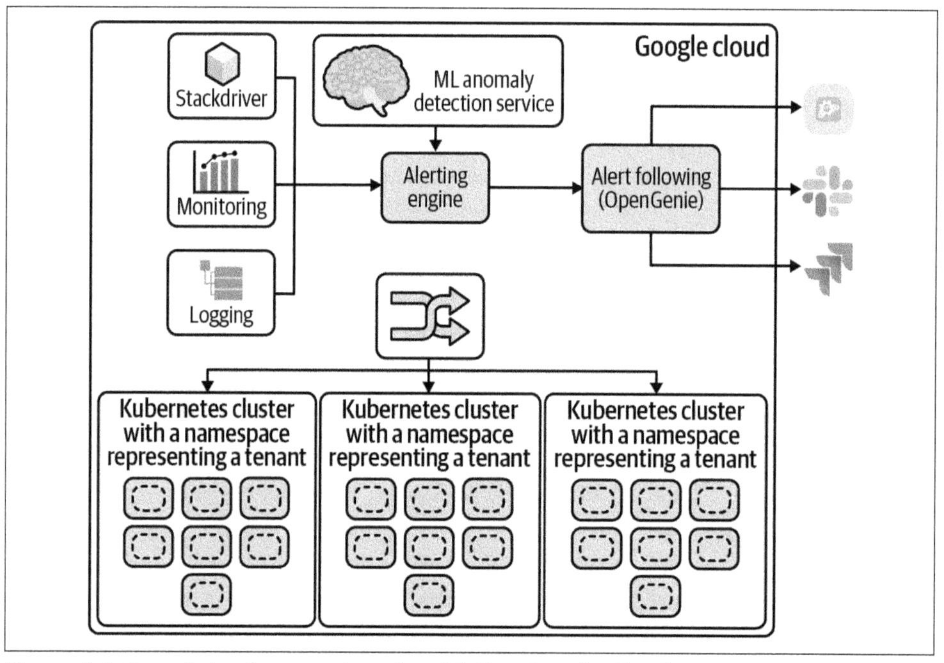

Figure 6-3. Sample implementation of an SOC in Google Cloud

Figure 6-3 shows a set of Kubernetes clusters running in Google Cloud with a namespace representing a tenant. Each tenant cluster can be deployed in Google Kubernetes Engine (GKE) or as an upstream Kubernetes cluster in Google Cloud. There is an ingress representing how the service is accessed by external entities. The details of the workload deployment and provisioning are omitted from the figure, as we want to focus on how you can secure the service. In order to secure the service, you need logging and monitoring and alerting. This can be achieved by using the Google Cloud operations suite (*https://oreil.ly/9rTas*), which provides capabilities to support logging and monitoring and alerting. In case you are using GKE, Google Cloud's blog (*https://oreil.ly/QCGa0*) describes how to leverage these services to detect and manage alerts for your Kubernetes clusters. As mentioned before, you need to leverage ML for baselining and to improve the quality of alerts. Google offers a set of ML services known as AI Hub (*https://oreil.ly/ICHTe*). Note you still need to build ML models that are relevant and effective for your SaaS service (see the example ML jobs earlier in this chapter). You can then use well-known tools like OpsGenie (*https://oreil.ly/gNSbr*) to route alerts for alert management to SIEM, Slack, PagerDuty, JIRA, and other tools. These alerts will then trigger the remediation workflows as defined by the security team. Note we have used Google Cloud as an example, but you can use the previously mentioned approach to build an SOC for AWS and Azure. These cloud providers also have a similar set of services available to users.

The previously mentioned approach is very effective in case you are using only one cloud provider and do not have any workloads running on-premise or in other cloud provider environments. Also, all the services mentioned earlier will increase the cost of deployment, and you also need to delegate some of your DevOps/DevSecOps resources to implement and manage these services. Therefore, we recommend that you build your SOC using a tool that is agnostic to any cloud provider and can be used across cloud providers/on-premise environments.

Figure 6-4 shows how you can replace some of the cloud provider–specific components and create an SOC using a Kubernetes-native observability and security platform. You can build the platform yourself or you can choose to use products that offer these platforms, such as Datadog, VMware, and Calico Enterprise. When you choose products, keep in mind the concepts covered in the previous section about alerting, and ensure that the platform supports integrations to your remediation/management systems.

Now that we have reviewed how you build an SOC that is effective for your Kubernetes cluster, let's review another application of observability to secure your Kubernetes cluster.

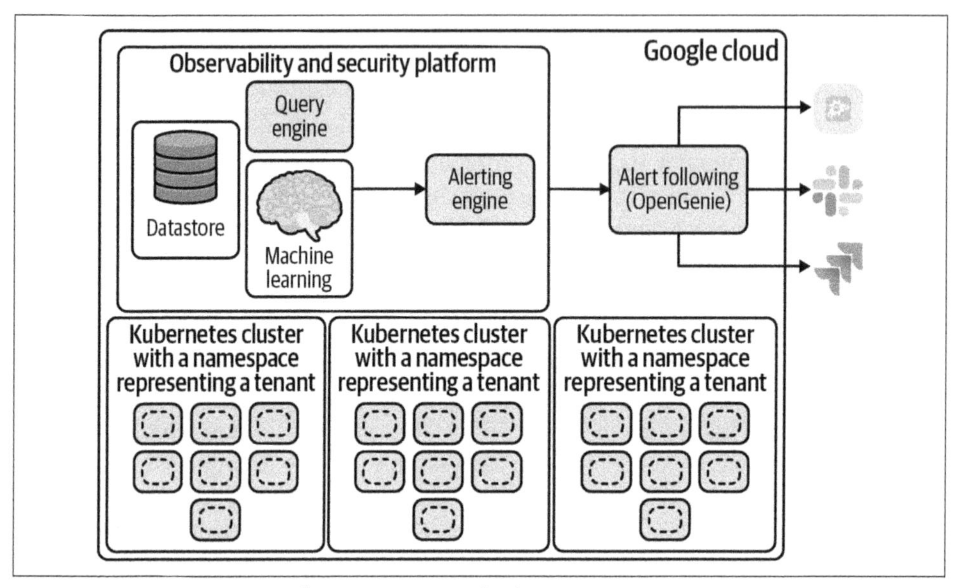

Figure 6-4. SOC using a Kubernetes-native platform

User and Entity Behavior Analytics

User and entity behavior analytics (UEBA) is an area where you use ML- and AI-based techniques to profile the behavior of a user or an entity (e.g., a pod, service, or deployment) over time and detect anomalous behavior by the user/entity. Microsoft Azure offers UEBA as a part of its cloud platform. Microsoft Azure's blog post, Identify Advanced Threats with User and Entity Behavior Analytics (UEBA) in Azure Sentinel" (*https://oreil.ly/FDspV*), and an excellent resources that describes how you can use UEBA for security use cases. Note that anomalous behavior by an entity is not always suspicious behavior; you need to map the behavior to frameworks like the MITRE attack framework or other indicators of compromise to confirm it is a security issue.

Let's take a simple example of how you can implement UEBA for an entity in Kubernetes, such as a service.

Figure 6-5 shows a service in your Kubernetes cluster and the various interactions of the service we will consider when we profile the behavior of the service. As a part of its normal operation, the service will interact with the Kubernetes API server and the Kubernetes datastore. In addition, it will interact with the ingress resource to communicate with entities external to the cluster and use cluster networking to interact with other entities inside the cluster. The service will also use the DNS service in the cluster for its operation.

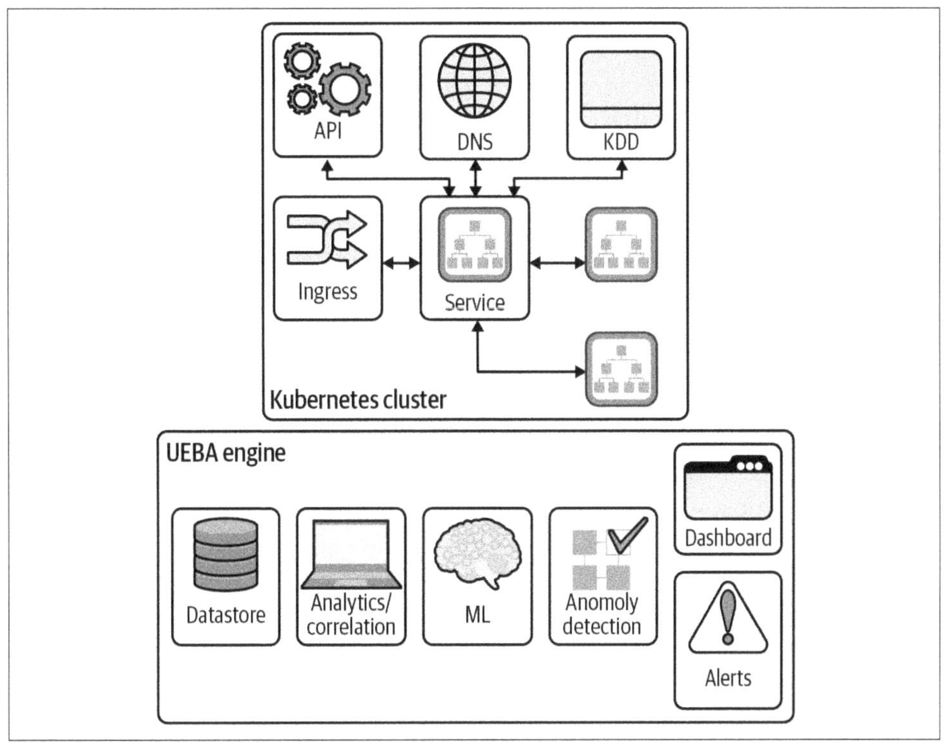

Figure 6-5. Profiling the behavior of a Kubernetes service

In order to build a profile for the service, we would need to consider the following aspects of the service. These are known as features in machine learning.

- Service composition (number of endpoints such as pods, RBAC, policies)
- Filesystem activity, process information, and system call activity for the service
- Service accounts associated with the service
- Service life cycle operations (e.g., create, delete, scale up/down)
- Traffic to and from the service (network, application)
- DNS activity by pods in the service

The UEBA engine shown in Figure 6-5 will collect logs from various data sources (network flow logs, application flow logs, Kubernetes audit logs, DNS activity logs, process information, filesystem, syscall activity logs) and store them in the datastore. These logs are aggregated and correlated by the analytics engine to generate correlated logs for the service across various features.

The machine learning engine uses a complex model to baseline the behavior of the service across various features. This is an advanced concept in machine learning where the model considers each feature and interactions between features, among other things, to build a profile for the service. This is best implemented by your data science team. This profile is then used to predict anomalies and generate alerts for deviations. There is a dashboard to allow SOC operators to review analyzed data and use it for forensics or to hunt threats. Please note a security and observability platform built using the concepts described in Chapter 5 will help build an effective UEBA system.

UEBA is an advanced technique and is complex to implement, but it is a very effective way to quickly find out which entities in your cluster are potentially vulnerable. This makes SOC operation very efficient and scalable. Once your deployment scales up to several clusters (50+), it is not practical to use alerts/manual reviews of dashboards to find real issues. UEBA will alert you to entities that are abnormal and need immediate attention.

Conclusion

When you think about how you can use observability to help secure your cluster, please consider the following:

- The alerting system you use must be Kubernetes-native and must support baselining using ML so that you do not have to manually define thresholds for various features.
- It is recommended that you consider a Kubernetes-native platform that works across your cloud and on-premise deployments to build your SOC.
- UEBA is an advanced concept and is complex to implement, but it can be very effective in securing a Kubernetes cluster.

CHAPTER 7

Network Policy

In this chapter, we will describe network policy and discuss its importance in securing a Kubernetes cluster. We will review various network policy implementations and tooling to support network policy implementations. We will also cover network policy best practices with examples.

What Is Network Policy?

Network policy is the primary tool for securing a Kubernetes network. It allows you to easily restrict the network traffic in your cluster so only the traffic that you want to flow is allowed.

To understand the significance of network policy, let's briefly explore how network security was typically achieved prior to network policy. Historically in enterprise networks, network security was provided by designing a physical topology of network devices (switches, routers, firewalls) and their associated configuration. The physical topology defined the security boundaries of the network. In the first phase of virtualization, the same network and network device constructs were virtualized in the cloud, and the same techniques for creating specific network topologies of (virtual) network devices were used to provide network security. Adding new applications or services often required additional network design to update the network topology and network device configuration to provide the desired security.

In contrast, the Kubernetes network model defines a "flat" network in which, by default, every pod can communicate directly with all other pods in the cluster. This approach massively simplifies network design and allows new workloads to be scheduled dynamically anywhere in the cluster with no dependencies on the network design.

In this model, rather than network security being defined by network topology boundaries, it is defined using network policies that are independent of the network topology. Network policies are further abstracted from the network by using label selectors as their primary mechanism for defining which workloads can talk to which workloads, rather than IP addresses or IP address ranges.

Network policy enforcement can be thought of as each pod being protected by its own dedicated virtual firewall that is automatically programmed and updated in real time based on the network policy that has been defined. Figure 7-1 shows network policy enforcement at a pod using its dedicated virtual firewall.

Figure 7-1. Pod secured by a virtual firewall

Why Is Network Policy Important?

In an age where attackers are becoming more and more sophisticated, network security as a line of defense is more important than ever.

While you can (and should) use firewalls to restrict traffic at the perimeters of your network (commonly referred to as north-south traffic), their ability to police Kubernetes traffic is often limited to a granularity of the cluster as a whole, rather than to specific groups of pods, due to the dynamic nature of pod scheduling and pod IP addresses. In addition, the goal of most attackers once they gain a small foothold inside the perimeter is to move laterally (east-west) to gain access to higher-value targets, which perimeter-based firewalls can't police against. With application architectures evolving from monoliths to microservices, the amount of east-west traffic, and therefore attack surface for lateral movement, is continuing to grow.

Network policy, on the other hand, is designed for the dynamic nature of Kubernetes by following the standard Kubernetes paradigm of using label selectors to define groups of pods, rather than IP addresses. And because network policy is enforced within the cluster itself, it can secure both north-south and east-west traffic.

Network policy represents an important evolution of network security, not just because it handles the dynamic nature of modern microservices, but because it empowers dev and DevOps engineers to easily define network security themselves, rather than needing to learn low-level networking details. Network policy makes it easy to define intent, such as *only this microservice gets to connect to the database*, write that intent as code (typically in *.yaml* files), and integrate authoring of network policies into Git workflows and CI/CD processes.

Network Policy Implementations

Kubernetes defines a standard network policy API, so there's a base set of features you can expect on any cluster. But Kubernetes itself doesn't do anything with the network policy other than store it. Enforcement of network policies is delegated to network plug-ins, allowing for a range of implementations. Most network plug-ins support the mainline elements of Kubernetes network policies, though many do not implement every feature of the specification. It's worth noting that most implementations are coupled with the network plug-in's specific pod networking implementation. However, some network policy implementations can enforce network policy on top of a variety of different pod networking plug-ins. Figure 7-2 shows network policies stored in the Kubernetes datastore being used by network plug-ins for enforcement.

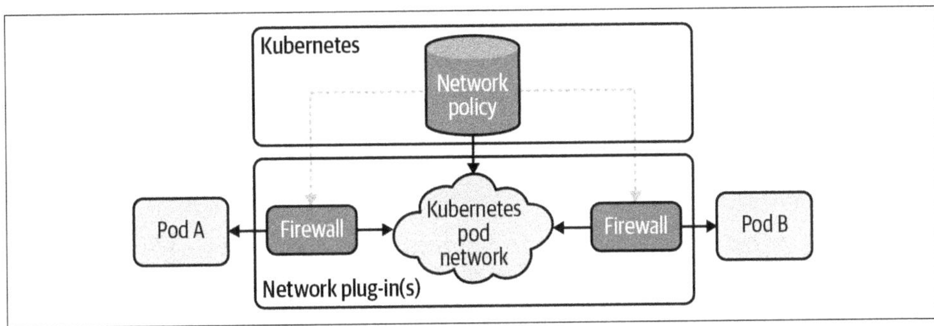

Figure 7-2. Network policy that is stored in Kubernetes enforced by network plug-ins

There are a number of networking and network policy implementations to choose from, as shown in Figure 7-3.

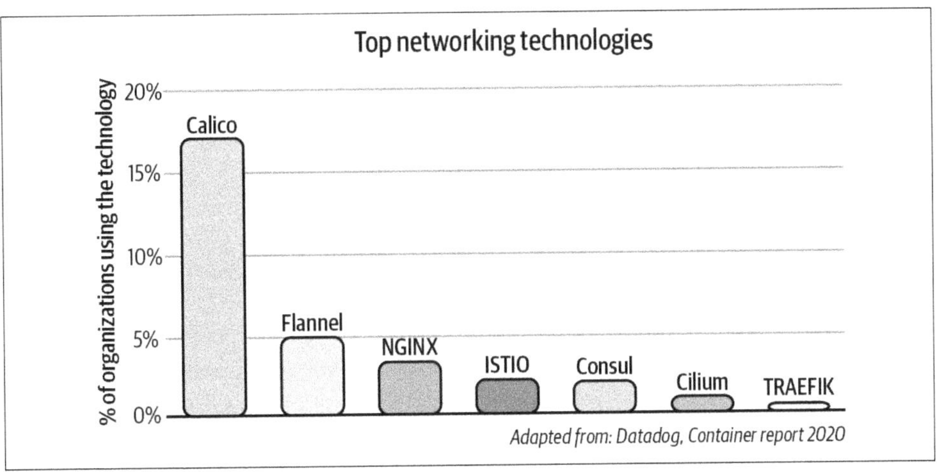

Figure 7-3. Adoption of top networking technology implementations

No matter what network policy implementation you choose, we recommend using one for the following reasons:

- It implements the complete Kubernetes network policy specification.
- In addition to supporting the Kubernetes network policy specification, its own policy model provides additional capabilities, which can be used alongside Kubernetes network policies to support additional enterprise security use cases.
- A few network plug-ins, like Weave Net, Kube-router, and Calico, can enforce network policy on top of their own rich set of networking capabilities, or on top of several other networking options, including the network plug-ins used by Amazon's Elastic Kubernetes Service (EKS), Azure Kubernetes Service (AKS), and Google Kubernetes Engine (GKE). This makes them a particularly strong choice as part of a multicloud strategy, because it gives you the flexibility to select the best networking for your environment from a broad range of options with the same rich set of network policy features available across all environments.
- The network policy can be applied to host endpoints/interfaces, allowing the same flexible policy model to be used to secure Kubernetes nodes or noncluster hosts/VMs.
- It supports network policy that is enforced both at the network/infrastructure layer and at the layers above, including supporting L5–L7 match criteria with its policy rules such as HTTP methods and paths. The multiple enforcement points help protect your infrastructure from compromised workloads and protect your workloads from compromised infrastructure. It also avoids the need for dual provisioning of security at the application and infrastructure layers, or having to learn different policy models for each layer.
- It needs to be production-grade, which means it must perform very well in clusters of any size, from single-node clusters to several-thousand-node clusters.
- It provides the ability for enterprises to add new capabilities and serves as a building block for an enterprise-grade Kubernetes network security solution.

Network Policy Best Practices

In this section we'll explore how to implement network policy with examples and cover best practices for implementation. The following examples use the Calico network policy schema, which extends the Kubernetes network policy schema. We're using these examples due to our familiarity with Calico network policy, but these best practices can be implemented with other available network policy models as well.

Ingress and Egress

When people think about network security, the first thought is often of how to protect your workloads from north-south external attackers. To help defend against this, you can use network policy to restrict ingress traffic to any pods that are reachable from outside the cluster.

However, when an attacker does manage to find a vulnerability, they often use the compromised workload as the place from which to move laterally, probing the rest of your network to exploit additional vulnerabilities that give them access to more valuable resources or allow them to elevate privileges to mount more powerful attacks or exfiltrate sensitive data.

Even if you have network policies to restrict ingress traffic on all pods in the cluster, the lateral movement may target assets outside of the cluster, which are less well protected. Consequently, the best practice is to always define both ingress and egress network policy rules for every pod in the cluster.

While this doesn't guarantee an attacker cannot find additional vulnerabilities, it does significantly reduce the available attack surface, making the attacker's job much harder. In addition, if combined with suitable alerting of policy violations, the time to identify that a workload has been compromised can be massively reduced. To put this into perspective, in the 2020 IBM Cost of a Data Breach Report, IBM reported that on average it took enterprises 207 days to identify a breach, and a further 73 days to contain it! With correctly authored network policies and alerting of violations, the breach can be prevented or reduced potentially to minutes or seconds and even opens the possibility of automated responses to quarantine the suspect workload if desired.

Not Just Mission-Critical Workloads

Best practice already recommends ensuring every pod has a network policy that restricts its ingress and egress traffic. What this means is that when you are thinking about how to protect your mission-critical workloads, you really need to be protecting all workloads. If you don't, then some seemingly unimportant, innocuous workload could end up being used as the base for attacks across the rest of your network, ultimately leading to the downfall of your most critical workloads.

Policy and Label Schemas

One of the strengths of Kubernetes labels and network policies is the flexibility in how you can use them. However, as a result there are often multiple different ways of labeling and writing policies that can achieve the same particular goal. So another best practice is to consider standardizing the way you label your pods and write your network policies using a consistent schema or design pattern. This can make authoring and understanding the intent of each network policy much more straightforward, especially if your clusters are hosting a large number of microservices.

For example, you might say every pod will have an "app" label that identifies which microservice it is, and every pod will have a single network policy applied to it using that app label, with the policy defining ingress and egress rules for the microservices it is expected to interact with, again using the app label:

```
apiVersion: projectcalico.org/v3
kind: NetworkPolicy
metadata:
  name: back-end-policy
  namespace: production
spec:
  selector: app == 'back-end'
  ingress:
  - action: Allow
    protocol: TCP
    source:
      selector: app == 'front-end'
    destination:
      ports:
        - 80
  egress:
  - action: Allow
    protocol: TCP
    destination:
      selector: app == 'database'
      ports:
        - 80
```

Or you might decide to use permission-style labels in the policy rules so that rather than listing the microservices that are allowed to access each service in its ingress rules, any microservice that has the permission label is allowed:

```
apiVersion: projectcalico.org/v3
kind: NetworkPolicy
metadata:
  name: database-policy
  namespace: production
spec:
  selector: app == 'database'
  ingress:
  - action: Allow
    protocol: TCP
    source:
      selector: database-client == 'true'
    destination:
      ports:
        - 80
  egress:
  - action: Deny
```

This could make it easier for individual microservice teams to author their own network policies without needing to know the full list of other microservices that must consume the service.

There are plenty of other ways you could go about it, and there is no right or wrong here. But taking the time to define how to approach labeling and defining network policies up front can make life significantly easier in the long run.

If you are not sure which approach will work best for you, then a simple "app" approach is a good place to start. This can always be expanded later to include the ideas of permission-style labels for microservices that have a lot of clients if maintaining the policy rules becomes time-consuming.

Default Deny and Default App Policy

The Kubernetes network policy specification allows all ingress pod traffic, unless there is one or more network policy with an ingress rule that applies to the pod, and then only the ingress traffic that is explicitly allowed by the policies is allowed. And likewise for egress pod traffic. As a result, if you forget to write a network policy for a new microservice, it will be left unsecured. And if you forget to write both ingress and egress rules for the microservice, then it will be left partially unsecured.

Given this, a good practice is to put in place a "default deny policy" that prevents any traffic that is not explicitly allowed by another network policy. The way this is normally done is to have a policy that specifies it applies to all pods, with both ingress and egress rules, but does not explicitly allow any traffic itself. As a result, if no other network policy applies that explicitly allows the traffic, then the traffic will be denied:

```
apiVersion: networking.k8s.io/v1
kind: NetworkPolicy
metadata:
  name: default-deny
  Namespace: my-namespace
spec:
  podSelector: {}
  policyTypes:
  - Ingress
  - Egress
```

Since Kubernetes network policy is namespaced, the network policy mentioned previously needs repeating for each namespace and ideally needs to be built into the standard operating procedure for provisioning new namespaces in the cluster. Alternatively, some network policy implementations go beyond Kubernetes network policy and provide the ability to specify cluster-wide network policies (that are not limited to a single namespace). The following example shows how to create a policy that switches the whole cluster to default deny behavior, including any namespaces that are created in the future:

```
apiVersion: projectcalico.org/v3
kind: GlobalNetworkPolicy
metadata:
  name: default-deny
spec:
  selector: all()
  types:
  - Ingress
  - Egress
```

However, it's worth noting that this policy applies to all pods, not just application pods, including control plane pods for Kubernetes. If you do not have the right network policies in place or failsafe ports configured before you create such a policy, you can break your cluster in pretty bad ways.

A much less high-stakes best practice is to define a network policy that applies only to pods, excluding control plane pods. As well as triggering default deny behavior, this policy can include any rules that you want to apply to all application pods. For example, you could include a rule that allows all application pods to access kube-DNS. This helps simplify any per-microservice policies that need writing so they can focus solely on the desired per-microservice specific behaviors:

```
apiVersion: projectcalico.org/v3
kind: GlobalNetworkPolicy
metadata:
  name: default-app-policy
spec:
  namespaceSelector: has(projectcalico.org/name) &&
projectcalico.org/name not in {"kube-system", "calico-system"}
  types:
  - Ingress
  - Egress
  egress:
    - action: Allow
      protocol: UDP
      destination:
        selector: k8s-app == "kube-dns"
        ports:
        - 53
```

As this policy deliberately excludes control plane components, to secure the control plane you can write specific policies for each control plane component. It is best to do any such policy creation at cluster creation time before the cluster is hosting workloads, since getting these policies wrong can leave your cluster in a broken state that might result in a significant production outage. In addition, it is highly recommended you always make sure you have in place the correct failsafe ports for the network plug-in you are using before you start trying to create any policies for the control plane.

Policy Tooling

In this section we explore tools at your disposal to effectively add network policies to your Kubernetes cluster.

Development Processes and Microservices Benefits

One of the advantages of network policy compared with traditional network security controls is that defining network policy does not require networking or firewall expertise. Network policies use the same concepts and paradigms as other Kubernetes resources. In theory, any team that is familiar with deploying microservices in Kubernetes can easily master network policies. As a result, network policy represents an opportunity to adopt a shift-left philosophy for network security, where network security is defined earlier in the development cycle, rather than being defined late in the process. This is a great opportunity for the security and development teams to collaborate to secure your Kubernetes cluster.

At the same time, many organizations are moving from monolith application architectures to microservice architectures, often with one of the goals being to increase development and organizational agility. In such an approach, each microservice is typically maintained by a single development team, with that team having significant expertise on the microservice, but not necessarily the whole of the application that the microservice is a part of. The move to microservices complements the shift-left opportunity of network policy. The team responsible for the development of a microservice normally has a good understanding of which other microservices it consumes and depends on. They may also have a good understanding of which microservices consume their microservice.

When coupled with a well-defined, standardized approach to policy and label schemas, this puts them in a strong position to implement network policies for their microservice as part of the development of the microservice. In this model, network policy is treated as code built into and tested during the development process, just like any other critical part of a microservice's code.

An equally valid approach is to have development teams focus purely on the internals of the microservices they are responsible for and leave responsibility for operating the microservices with DevOps teams. However, the same ideas still apply. Such a DevOps team typically needs a good understanding of the dependencies between the microservices they are responsible for operating in order to manage the operation of the application and life cycle of the microservices. Network security can be defined as code by the DevOps team and tested just like they would any other operational code or scripts they develop before using in production.

The reality today, of course, is that many organizations are some way off from achieving this nirvana of microservices, agility, and shift-left security. Network security may come much later in the organization's processes, or even as an afterthought on a system already in production. In such scenarios, defining network policies may be significantly more challenging, and getting network policies wrong could have significant production impacts. The good news is that there are a range of tools to help with network policy life cycle management to make this easier, including policy recommendations, policy impact previews, and policy staging/audit modes.

Policy Recommendations

Policy recommendation tools are a great help in scenarios where the team responsible for network security does not have a good, confident understanding of all the network dependencies between the applications or microservices they need to secure. These tools also help you get started with authoring network policies the right way, and make the creation of network policy significantly easier than writing it by hand.

The way recommendation tools usually work is to analyze the network traffic to and from each microservices over a period of time. This means to get recommendations, the microservice needs to be running in production, or a staging or test environment that accurately reflects the production interactions between the microservice and the rest of the application.

There are many policy recommendation tools available to choose from, often with varying levels of sophistication, degrees of Kubernetes awareness, and policy schema approaches. It is recommended that you use a Kubernetes-aware policy recommendation engine built into your network policy solution.

Policy Impact Previews

Policy impact preview tools provide a way of sanity-checking a network policy before it is applied to the cluster. Like policy recommendations, this is usually done by analyzing the cluster's historical network traffic over a period of time and calculating which network flows would have been impacted by the new policy. An example is to identify any flows that were previously allowed that would now be denied, and any flows that were previously denied that would now be allowed.

Policy impact previews are a great help in any scenarios where you are not relying completely on policy recommendations. For example, this could be if you are defining network policies by hand or modifying a policy recommendation to align with a particular standardized approach to policy and label schemas. Even if the team defining the network policy for a microservice has high confidence in their understanding of the microservice's network dependencies, policy impact previews can be invaluable to help catch any accidental mistakes, such as hard-to-spot typos, that might significantly impact legitimate network traffic.

Policy impact preview tools are less common than policy recommendations. It is very useful to use a tool that provides a visual representation of the impact based on analysis of the flow log data it collects over any desired time period. This will help in reducing issues due to incorrectly authored policies or outages due to operator error.

Policy Staging and Audit Modes

Even less common than policy impact previews, but potentially even more valuable, is support for policy staging, sometimes called policy audit mode.

Policy staging allows network policy changes to be applied to the cluster without impacting network traffic. The staged policy then records the full details of all the flows it would have interacted with, without actually impacting any of the flows. This is incredibly useful in scenarios where a policy impact preview of an individual policy against historical data may be overly simplistic given the complexity of the applications running in the cluster. For example, this could be if multiple interdependent policies need to be updated in unison, or if there's a desire to monitor the policy impact with live rather than historical network flows.

In order to make the task of authoring effective network policies less daunting, you need to use policy recommendations and then stage policies to understand the impact of the policy before you promote it for enforcement. This cycle of policy recommendation (based on historical network flows), followed by staging (applying policies to current and future network flows), followed by making desired adjustments and then finally enforcing the policy is the best way to ensure the policy change would do exactly what you want.

Conclusion

In this chapter we discussed the importance of network policy and various network policy implementations and tooling to help you with implementation. The following are some key aspects of network policy:

- Network policy should be used to secure a Kubernetes network, and it complements the firewalls that are implemented at the perimeter of your cluster.
- It is recommended that you choose a Kubernetes-aware implementation that extends the basic Kubernetes network policy.
- There are a lot of network policy implementations that offer tooling to help with the implementation of network policy in a Kubernetes cluster.

CHAPTER 8
Managing Trust Across Teams

In the previous chapter we explored how network policy represents an opportunity to adopt a shift-left philosophy for network security, where security is defined by teams earlier in the development cycle rather than being defined and maintained by a security team late in the process. This approach can bring a lot of benefits, but to be viable, there needs to be a corresponding degree of trust and split of responsibilities between the teams involved.

In most organizations, it's not practical to shift 100% of the responsibility for security all the way to the left, with all other teams (platform, network, and security) washing their hands of any responsibility for security. So for example, while the responsibility for lower-level details of individual microservice security may be shifted left, the security team may still be responsible for ensuring that your Kubernetes deployment has a security posture that meets internal and external compliance requirements.

Some enterprises handle this by defining internal processes, for example, to ensure the security team reviews all security changes before they are applied. The downside of this approach is it can reduce agility, which is at odds with one of the motivations for shifting left, that being to increase agility.

Fortunately, there are various types of guardrails that can be put in place across a Kubernetes environment that reduce the need for these kinds of traditional process controls. In this chapter we will explore some of these capabilities and how they can be used to control the degree of trust being delegated from one team to another in the context of a shift-left approach to security.

Role-Based Access Control

Kubernetes role-based access control (RBAC) is the primary tool for defining the scope of what individual users or groups of users are permitted to do in a Kubernetes cluster. RBAC permissions are defined using roles and granted to users or groups of users via role bindings. Each role includes a list of resources (specified by resource type, cluster-wide, within a namespace, or even a specific resource instance) and the permissions for each of the resources (e.g., get, list, create, update, delete, etc.).

Many Kubernetes resources are namespaced, including deployments, daemonsets, pods, and Kubernetes network policies. This makes the namespace an ideal trust boundary between teams. There are no set rules for how to use namespaces, but one common practice is to use a namespace per microservice. RBAC can then be used to grant permission to manage the resources in the namespace to the team responsible for operating the corresponding microservice.

If security has been shifted left, this would normally include permissions to manage the network policies that apply to the microservice, but not to manage any network policies that apply to microservices they are not responsible for.

If default deny–style best practices are being followed for both ingress and egress traffic, then the team cannot forget to write network policies, because the microservice will not work without them. In addition, since other teams will have defined equivalent network policies covering both ingress and egress traffic for the microservices they are responsible for, traffic is allowed between two microservices only if both teams have specified network policy that says the traffic is allowed. This further controls the degree of trust being delegated to each team.

Of course, depending on the degree to which security has been shifted left, the responsibility for defining network policies may fall to a different team than the team responsible for operating the microservice. Again, Kubernetes RBAC can be used to easily reflect this split of responsibilities.

Limitations with Kubernetes Network Policies

There are a couple of limitations it is worth being aware of when using RBAC with Kubernetes network policies in a shift-left environment:

- Default deny–style policies need to be created per namespace at the time the namespace is provisioned. The team responsible for defining network policies for the microservice would also have the ability to modify or delete this default policy if they wanted to.
- Network policies are IP-based, and you cannot use fully qualified domain names (FQDNs). This can be a limitation especially when defining policies to resources external to the cluster.

- Kubernetes RBAC controls access to resources but does not constrain the contents of resources. Of particular relevance in the context of network policies are pod labels, since these are used as the primary mechanism for identifying other microservices in network policy rules. So for example, if one team has written a network policy for their microservice with a rule allowing traffic to it from pods with a particular label, then in theory any team with permission to manage pods could add that label to their pods and get access to the microservice. This exposure can be reduced by always using namespace sectors within policy rules and being selective as to which teams have permissions to change namespace labels.

If standardized policy and label schemas have been defined and the teams are trusted to follow them, then these limitations are more of a theoretical rather than practical issue. However, for some organizations, they may represent genuine issues for their security needs. These organizations may therefore want to leverage additional capabilities beyond Kubernetes RBAC and Kubernetes network policies. In particular, they might consider:

- Richer network policy implementations that support additional network policy types, match criteria, and non-namespaced network policies, which open up more options for how to split responsibilities and RBAC across teams
- Admission controllers to enforce controls on a per-field level within resources, for example to ensure a standardized network policy and label schemas are followed, including limiting teams to using particular labels

We will now review network policy implementations that extend the Kubernetes network policy and how you can use the same to manage trust.

Richer Network Policy Implementations

Some network policy implementations support both Kubernetes network policies and their own custom network policy resources that can be used alongside or instead of Kubernetes network policies. Depending on the implementation, these may open up additional options for how to split responsibilities and use RBAC across teams. There are vendors that offer richer network policy implementations that support the Kubernetes network policy and add more features (e.g., Weave Net, Kube-router, Antrea, Calico). We encourage you to review these and choose the best one that meets your needs. In this section we will look at the concrete example using Calico, as it is the most widely deployed container network plug-in.

Calico supports the Kubernetes network policy feature set, plus its own Calico network policy resources, which can be used alongside Kubernetes network policies. There are two types of Calico network policies, both under the projectcalico.org/v3 API group:

NetworkPolicy
: These policies are namespaced (just like Kubernetes network policies).

GlobalNetworkPolicy
: These policies apply across the whole of the cluster independent of namespace.

Both types of Calico network policy support a common set of capabilities beyond Kubernetes network policies, including:

- A richer set of match criteria than Kubernetes network policies, for example with the ability to match on Kubernetes service accounts.
- Explicit allow, deny, or log actions for policy rules, rather than Kubernetes network policy actions, which are implicitly always allow.
- Precedence ordering to define the evaluation order of the network policies if multiple policies apply to the same workload. (Note that if you are just using Kubernetes network policies, or Calico policies only with allow actions in them, then evaluation order doesn't make any difference to the outcome of the policies. However, as soon as there are any policy rules with deny actions, ordering becomes important.)

We want to mention that there are other network policy implementations that extend the Kubernetes network policy, like Antrea, which offers ClusterNetworkPolicy (similar to GlobalNetworkPolicy).

The following sample shows how you can implement network policies using Kubernetes RBAC. In the example you can control network access based on the labels assigned to a service account. In Kubernetes, pods have service accounts associated with them, and therefore pods can be identified by service accounts. You should use RBAC to control which users can assign labels to service accounts. The network policy in the example uses the labels assigned to service accounts to control network access. Pods with an intern service account can communicate only with pods with service accounts labeled `role == intern`:

```
apiVersion: projectcalico.org/v3
kind: NetworkPolicy
metadata:
  name: restrict-intern-access
  namespace: prod-engineering
spec:
  serviceAccountSelector: 'role == "intern"'
  ingress:
    - action: Allow
      source:
        serviceAccounts:
          selector: 'role == "intern"'
  egress:
    - action: Allow
```

```
    destination:
      serviceAccounts:
        selector: 'role == "intern"'
```

This way you can extend the concept of RBAC, which controls service account access to a Kubernetes resource for network access. It is a two-step process. RBAC is used to control label assignment to service accounts, and a label-based service account selector is used to control network access. These additional capabilities can be leveraged alongside Kubernetes network policies to more cleanly split responsibilities between higher-level cluster ops or security teams and individual microservice teams.

This could look like, for example:

- Giving the cluster ops or security team RBAC permissions to manage Calico network policies at the cluster-wide scope, so they can define basic higher-level rules that set the overall security posture of the cluster. For example, a default deny–style app policy (as discussed in Chapter 7) and policies can restrict cluster egress to specific pods.
- Giving each microservice team RBAC permissions to define Kubernetes network policies in the microservice's namespaces, so they can define their own fine-grained constraints for the microservices they are responsible for.

On top of this basic split in network policy RBAC permissions, the cluster ops or security team can delegate different levels of trust to each microservice team by defining rules using namespaces or service accounts labels, rather than simplifying matching on pod labels. For example, they could define policies to restrict cluster egress to specific pods using service account labels and give the individual microservice teams permissions to use, but not edit, any service accounts assigned to their namespace. Through this mechanism some microservice teams may be granted permission to selectively allow cluster egress from some of their pods, while not offering other teams the same permissions.

Figure 8-1 provides a simplified illustration of how these ideas can be combined.

While these capabilities are reasonably powerful, in some organizations the required split of responsibilities across teams may be more complex, particularly where there are more layers of teams. For example, a compliance team, security team, cluster ops team, and individual microservice teams all may have different levels of responsibility. One way to more easily meet these requirements is to use a network policy implementation that supports the notion of hierarchical network policies.

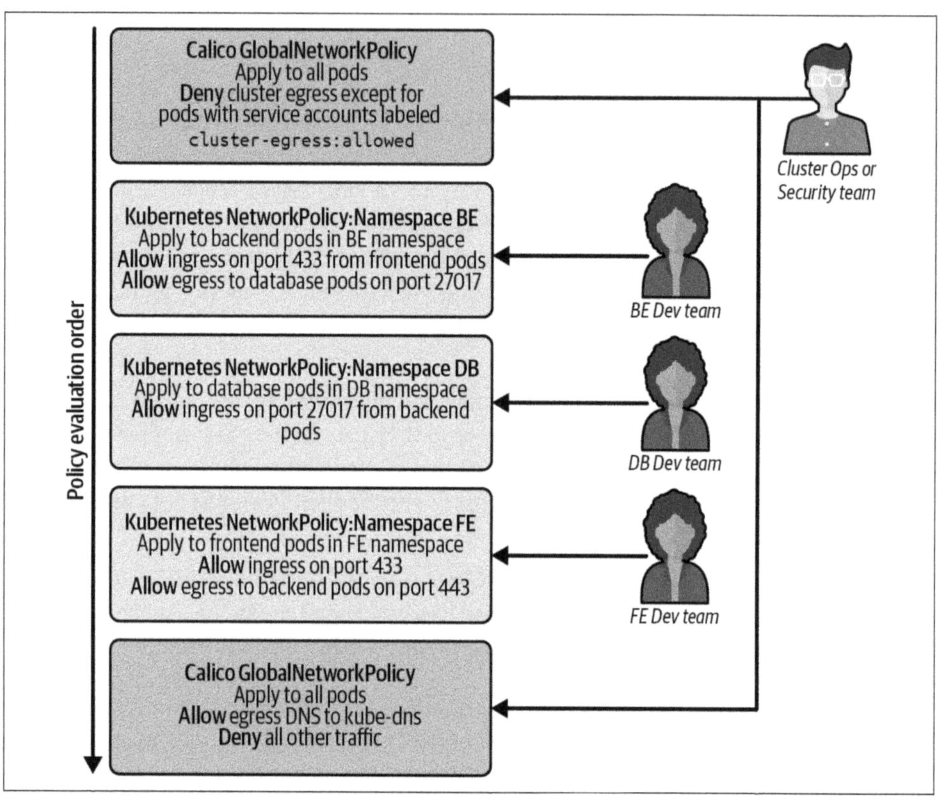

Figure 8-1. An example of implementing trust boundaries with network policy

There are some commercial implementations that support hierarchical network policy using policy tiers. A similar concept (hierarchical namespaces and policies) is also being discussed in the Kubernetes community. RBAC for each tier can be defined to restrict who can interact with the tier. In this model, network policies are layered in tiers, which are evaluated in a defined order, with as many tiers as required to match the organizational split of responsibilities. RBAC for each tier can be defined to restrict who can interact with the tier. The network policies in each tier can make allow or deny decisions (that terminate evaluation of any following policies) or pass the decision on to the next tier in the hierarchy to be evaluated against the policies in that tier.

Figure 8-2 provides a simplified illustration of how these capabilities can be used to split responsibilities across three distinct layers of responsibility within an organization.

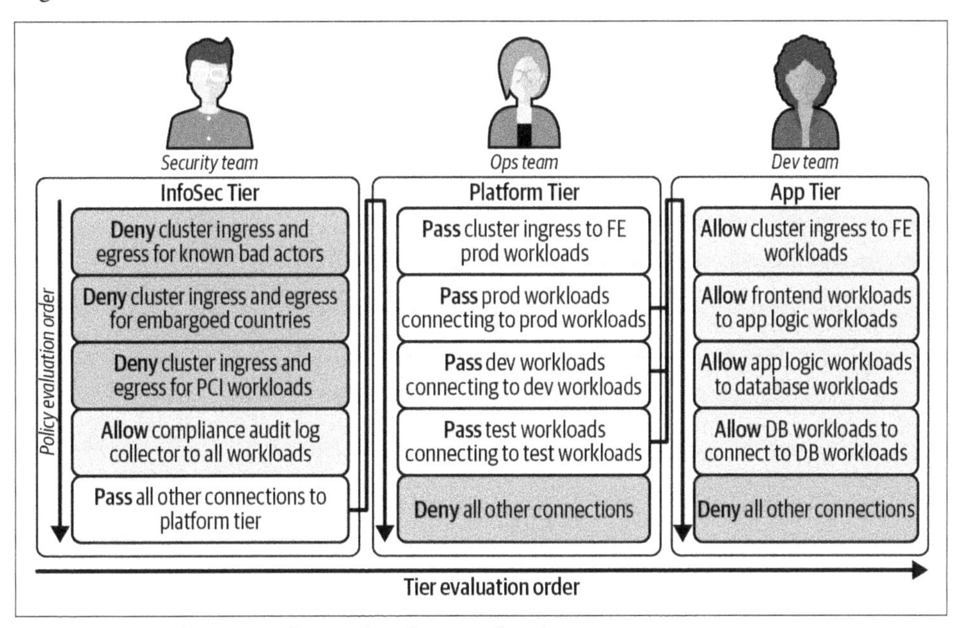

Figure 8-2. Implementing hierarchical network policies using tiers

Admission Controllers

We have already discussed the value of defining and following standardized network policy and label schemas. The approaches for splitting responsibilities between teams discussed earlier are oriented around resource- and namespace-level RBAC, with teams having freedom to whatever they want within the resource and namespace scopes they are allowed to manage. As such, they do not provide any guarantees that any such schemas are being followed by all teams.

Kubernetes itself does not have a built-in ability to enforce restrictions at this granular level, but it does support an Admission Controller API, which allows third-party admission controllers to be plugged into the Kubernetes API machinery to perform semantic validation of objects during create, update, and delete operations. You can additionally use admission controllers, also known as mutating admission controllers, for modifying objects that are admitted.

For example, in the context of implementing network policy, admission controllers can help with the following:

- Validate that network policies have both ingress and egress rules to comply with the best practices the organization is trying to follow.
- Ensure every pod has a specific set of labels to comply with the labeling standards the organization has defined.
- Restrict different groups of users to specific label values.

But admission controllers have security use cases beyond network policy too. For example, Kubernetes services include support for specifying an arbitrary IP address to be associated with the service using the service's ExternalIP field. Without some level of policing, this is a very powerful feature that could be used maliciously to intercept pod traffic to an IP address and redirect it to the Kubernetes service by anyone with RBAC permissions to create and manage Kubernetes services. Policing this with an admission controller might be essential depending on the level of trust within the teams involved.

There are a few options for admission controller implementations, depending on the skill sets and specific needs of the organization:

- Using a preexisting third-party admission controller that specializes in the specific controls the organization needs, if one exists
- Writing a custom admission controller optimized for the organization's needs
- Using a general-purpose admission controller with a rich policy model that can map to a broad range of use cases

For many scenarios, choosing a general-purpose admission controller gives a good balance of flexibility and coding complexity. For example, you might consider Kyverno, which has a policy engine specifically designed for Kubernetes, or an admission controller built around Open Policy Agent, where the policy model has flexible matching and language capabilities defined using Rego.

While admission controllers are very powerful, it is generally recommended to implement them only if you genuinely need them. For some organizations, using admission controllers in this way is overkill, given the levels of responsibility and trust across teams. For other organizations, they can be essential to meet internal compliance requirements, and the case for using them will be very clear.

Conclusion

Kuberentes security needs to be implemented by various teams and needs collaboration between teams. This chapter covered the following key concepts:

- You should use RBAC and network policy to define boundaries that will help you manage activities across teams.
- You can extend the concept of RBAC to control network access by leveraging service accounts in network policy to help you manage trust.
- Admission controllers help to control access and implement trust boundaries across various teams.
- Collaboration between the development, platform, and the security teams is important to implement security.

CHAPTER 9

Exposing Services to External Clients

In earlier chapters we explored how network policy is one of the primary tools for securing Kubernetes. This is true for both pod-to-pod traffic within the cluster (east-west traffic) and for traffic between pods and external entities outside of the cluster (north-south traffic). For all of these traffic types, the best practice is the same: Use network policy to limit which network connections are allowed to the narrowest scope needed, so the only connections that are allowed are the ones you expect and need for your specific applications or microservices to work.

In the case of pods that need to be accessed by external clients outside of the cluster, this means restricting connections:

- To the specific port(s) that the corresponding microservice is expecting incoming connections to
- From the specific clients that need to connect to the microservice

It's not uncommon for a particular microservice to be consumed just within the enterprise (whether on-prem or in a public cloud) by a limited number of clients. In this case the Kubernetes network policy rules ideally should limit incoming connections to just the IP addresses, or IP address range, associated with the clients. Even if a microservice is being exposed to the public internet (for example, exposing the frontend microservices for a publicly accessible SaaS or website), there are still cases where access may need to be restricted to some extent. For example, it may be a requirement to block access from certain geographies for compliance reasons, or it may be desirable to block access from known bad actors or threat feeds.

Unfortunately, how you go about implementing this best practice needs to include the consideration of which network plug-ins and Kubernetes primitives are used to expose the microservice outside the cluster. In particular, in some cases the original client source IP address is preserved all the way to the pod, which allows Kubernetes network policies to easily limit access to specific clients. In other cases the client source IP gets obscured by network address translation (NAT) associated with network load balancing, or by connection termination associated with application layer load balancing.

In this chapter we will explore different client source IP behaviors available across the three main options for exposing an application or microservice outside of the cluster: direct pod connections, Kubernetes services, and Kubernetes Ingress.

Understanding Direct Pod Connections

It's relatively uncommon for pods to be directly accessed by clients outside of the cluster rather than being accessed via a Kubernetes service or Kubernetes Ingress. However, there are scenarios where this may be desired or required. For example, some types of distributed data stores may require multiple pods, each with specific IP addresses that can be configured for data distribution or clients to peer with.

Supporting direct connections to pod IP addresses from outside of the cluster requires a pod network that makes pod IP addresses routable beyond the boundary of the cluster. This typically means using one of the following:

- A cloud provider network plug-in in public cloud clusters (e.g., the Amazon VPC CNI plug-in, as used by default in EKS)
- A network plug-in that can use BGP to integrate with an on-prem enterprise network (e.g., Kube-router, Calico CNI plug-in).

In addition to the underlying networking supporting the connectivity, the clients need a way of finding out the pod IP addresses. This may be done via DNS, explicit configuration of the client, or some other third-party service discovery mechanism.

From a security point of view, connections from clients directly to pods are straightforward: They have the original client source IP address in place all the way to the pod, which means network policy can easily be used to restrict access to clients with particular IP addresses or from particular IP address ranges.

Note though that in any cluster where pod IP addresses are routable beyond the boundary of the cluster, it becomes even more important to ensure network policy best practices are followed. Without network policy in place, pods that should only be receiving east-west connections could be accessed from outside of the cluster without the need for configuring a corresponding externally accessible Kubernetes service type or Kubernetes Ingress.

Understanding Kubernetes Services

Kubernetes services provide a convenient mechanism for accessing pods from outside of the cluster using services of type NodePort or LoadBalancer or by explicitly configuring an External IP for the service. By default, Kubernetes services are implemented by kube-proxy. Kube-proxy runs on every node in the cluster and is responsible for intercepting connections to Kubernetes services and load-balancing them across the pods backing the corresponding service. This connection handling has a well-defined behavior for when source IP addresses are preserved and when they are not, which we will look at now for each service type.

Cluster IP Services

Before we dig into exposing pods to external clients using Kubernetes services, it is worth understanding how Kubernetes services behave for connections originating from inside the cluster. The primary mechanism for service discovery and load balancing of these connections within the cluster (i.e., pod-to-pod connectivity) makes use of Kubernetes services of type Cluster IP. For Cluster IP services, kube-proxy is able to use destination network address translation (DNAT) to map connections to the service's Cluster IP to the pods backing the service. This mapping is reversed for any return packets on the connection. The mapping is done without changing the source IP address, as illustrated in Figure 9-1.

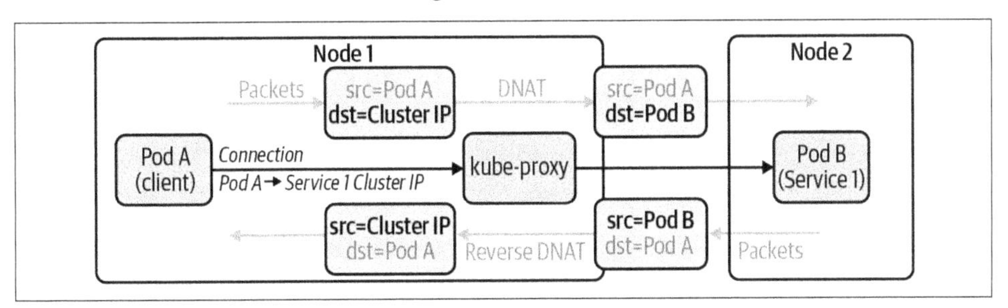

Figure 9-1. Network path for a Kubernetes service advertising Cluster IP

Importantly, the destination pod sees the connection has originated from the IP address of the client pod. This means that any network policy applied to the destination pod behaves as expected and is not impacted by the fact that the connection was load balanced via the service's Cluster IP. In addition, any network policy egress rules that apply to the client pod are evaluated after the mapping from Cluster IP to destination pod has happened. This means that network policy applied to the client pod also behaves as expected, independent of the fact that the connection was load balanced via the service's cluster IP. (As a reminder, network policy rules match on pod labels, not on service labels.)

Node Port Services

The most basic way to access a service from outside the cluster is to use a Kubernetes service of type NodePort. A node port is a port reserved on each node in the cluster through which the service can be accessed. In a typical Kubernetes deployment, kube-proxy is responsible for intercepting connections to node ports and load-balancing them across the pods backing each service.

As part of this process, NAT is used to map the destination IP address and port from the node IP and node port to the chosen backing pod and service port. However, unlike connections to cluster IPs, where the NAT maps only the destination IP address, in the case of node ports the source IP address is also mapped from the client IP to the node IP.

If the source IP address was not mapped in this way, then any response packets on the connection would flow directly back to the external client, bypassing the ability for kube-proxy on the original ingress node to reverse the mapping of the destination IP address. (It's the node that performed the NAT that has the connection tracking state needed to reverse the NAT.) As a result, the external client would drop the packets because it would not recognize them as being part of the connection it made to the node port on the original ingress node.

The process is illustrated in Figure 9-2.

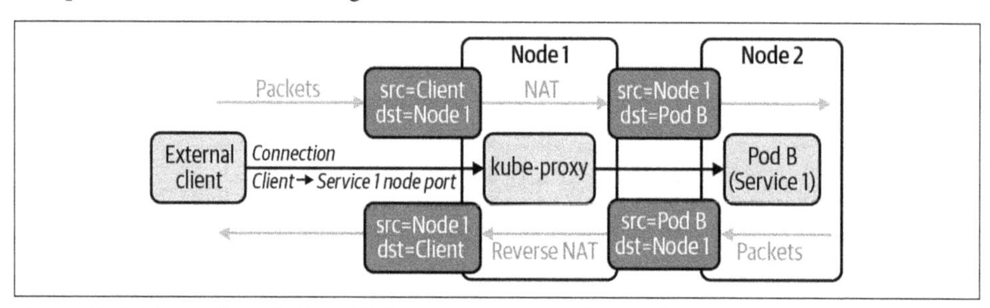

Figure 9-2. Network path for a Kubernetes service using node ports

Since the NAT changes the source IP address, any network policy that applies to the destination pod cannot match on the original client IP address. Typically this means that any such policy is limited to restricting the destination protocol and port and cannot restrict based on the external client's IP address. This in turn means the best practice of limiting access to the specific clients that need to connect to the microservice cannot easily be implemented with Kubernetes network policy in this configuration.

Fortunately, there are a number of solutions that can be used to circumvent the limitations of this default behavior of node ports:

- Configuring the service with `externalTrafficPolicy:local`
- Using a network plug-in that supports node-port-aware network policy extensions
- Using an alternative implementation for service load balancing in place of kube-proxy that preserves client source IP addresses

We will cover each of these later in this chapter. But before that, to complete our picture of how the default behavior of mainline Kubernetes services work, let's look at services of type LoadBalancer.

Load Balancer Services

Services of type LoadBalancer build on the behavior of node ports by integrating with external network load balancers. The exact type of network load balancer depends on which public cloud provider, or if on-prem, which specific hardware load balancer integration, is integrated with your cluster.

The service can be accessed from outside of the cluster via a specific IP address on the network load balancer, which by default will load-balance evenly across the nodes to the service's node port.

Most network load balancers are located at a point in the network where return traffic on a connection will always be routed via the network load balancer, and therefore they can implement their load balancing using only DNAT, leaving the client source IP address unaltered by the network load balancer, as illustrated in Figure 9-3.

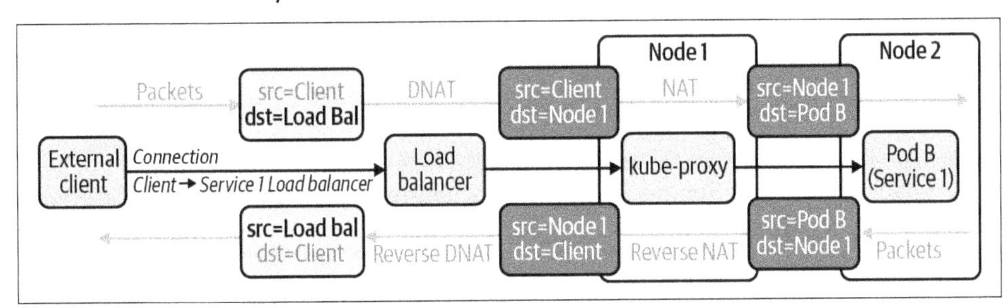

Figure 9-3. Network path for a Kubernetes service of type LoadBalancer

However, because the network load balancer is load-balancing to the service's node port, and kube-proxy's default node port behavior changes the source IP address as part of its load balancing implementation, the destination pod still cannot match on the original client source IP address. Just like with vanilla node port services, this in turn means the best practice of limiting access to the specific clients that need to connect to the microservice cannot easily be implemented with Kubernetes network policy in this configuration.

Fortunately, the same solutions that can be used to circumvent the limitations of the default behavior of services of type NodePort can be used in conjunction with services of type LoadBalancer:

- Configuring the service with `externalTrafficPolicy:local`
- Using a network plug-in that supports node-port-aware network policy extensions
- Using an alternative implementation for service load balancing in place of kube-proxy that preserves client source IP addresses

Let's look at each of those now.

externalTrafficPolicy:local

By default, the node port associated with a service is available on every node in the cluster, and services of type LoadBalancer load-balance to the service's node port evenly across all of the nodes, independent of which nodes may actually be hosting backing pods for the service. This behavior can be changed by configuring the service with `externalTrafficPolicy:local`, which specifies that connections should only be load balanced to pods backing the service on the local node.

When combined with services of type LoadBalancer, connections to the service are only directed to nodes that host at least one pod backing the service. This reduces the potential extra network hop between nodes associated with kube-proxy's normal node port handling. Perhaps more importantly, since each node's kube-proxy is only load-balancing to pods on the same node, kube-proxy does not need to perform source network address translation as part of the load balancing, meaning that the client source IP address is preserved all the way to the pod. (As a reminder, kube-proxy's default handling of node ports on the ingress node normally needs to NAT the source IP address so that return traffic flows back via the original ingress node, since that is the node that has the required traffic state to reverse the NAT.)

Network flow is illustrated in Figure 9-4.

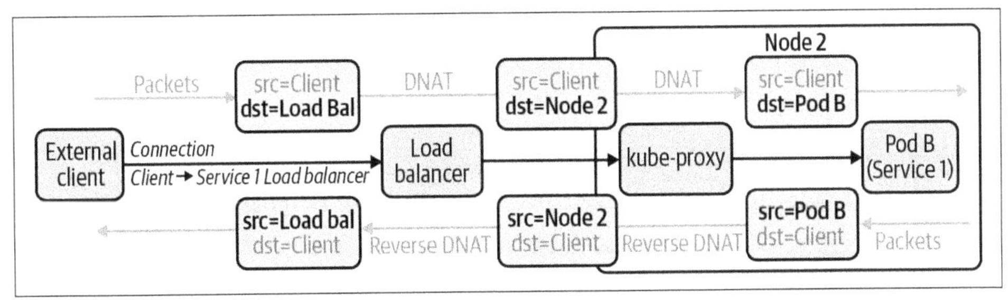

Figure 9-4. Network path for a Kubernetes service leveraging the optimization to route to the node backing the pod

As the original client source IP address is preserved all the way to the backing pod, network policy applied to the backing pod is now able to restrict access to the service to only the specific client IP addresses or address ranges that need to be able to access the service.

Note that not all load balancers support this mode of operation. So it is important to check whether this is supported by the specific public cloud provider, or if on-prem, the specific hardware load balancer integration, that is integrated with your cluster. The good news is that most of the large public providers do support this mode. Some load balancers can even go a step further, bypassing kube-proxy and load-balancing directly to the backing pods without using the node port.

Network Policy Extensions

Some Kubernetes network plug-ins provide extensions to the standard Kubernetes network policy capabilities, which can be used to help secure access to services from outside of the cluster.

There are many solutions that provide network policy extensions (e.g., Weave Net, Kuberouter, Calico). Let's look at Calico once again, as it's our area of expertise. Calico includes support for host endpoints, which allow network policies to be applied to the nodes within a cluster, not just pods within the cluster. Whereas standard Kubernetes network policy can be thought of as providing a virtual firewall within the pod network in front of every pod, Calico's host endpoint extensions can be thought of as providing a virtual firewall in front of every node/host, as illustrated in Figure 9-5.

In addition, Calico's network policy extensions support the ability to specify whether the policy rules applied to host endpoints apply before or after the NAT associated with kube-proxy's load balancing. This means that they can be used to limit which clients can connect to specific node ports, unencumbered by whatever load-balancing decisions kube-proxy may be about to make.

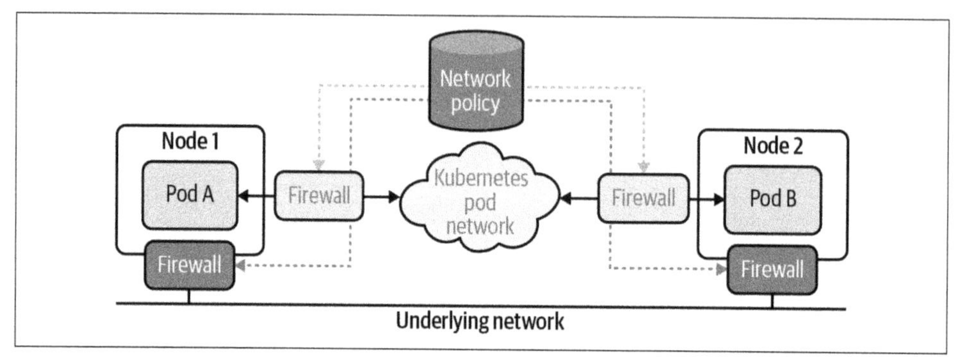

Figure 9-5. Virtual firewall using host endpoint protection

Alternatives to kube-proxy

Kube-proxy provides the default implementation for Kubernetes services and is included as standard in most clusters. However, some network plug-ins provide alternative implementations of Kubernetes services to replace kube-proxy.

For some network plug-ins, this alternative implementation is necessary because the particular way the plug-in implements pod networking is not compatible with kube-proxy's dataplane (which uses the standard Linux networking pipeline controlled by iptables and/or IPVS). For other network plug-ins, the alternative implementation is optional. For example, a CNI that implements a Linux eBPF dataplane will choose to replace kube-proxy in favor of its native service implementation.

Some of these alternative implementations provide additional capabilities beyond kube-proxy's behavior. One such additional capability that is relevant from a security perspective is the ability to preserve the client source IP address all the way to the back pods when load-balancing from external clients.

For example, Figure 9-6 illustrates how an eBPF-based dataplane implements this behavior.

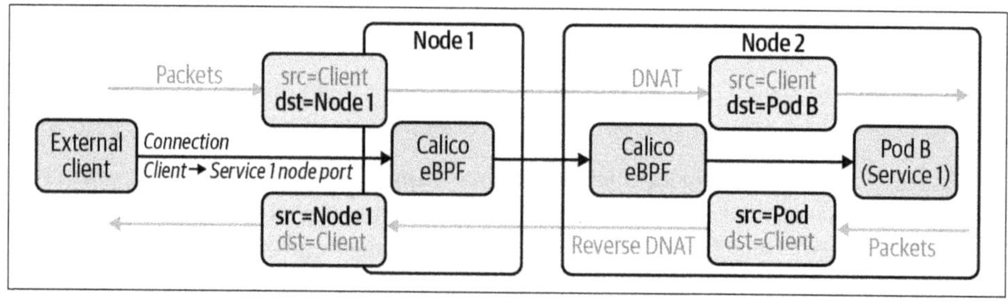

Figure 9-6. Network path for a Kubernetes service with an eBPF-based implementation

This allows the original client source IP address to be preserved all the way to the packing pod for services of type NodePort or LoadBalancer, without requiring support for `externalTrafficPolicy:local` in network load balancers or node selection for node ports. This in turn means that network policy applied to the backing pod is able to restrict access to the service to only the specific clients, IP addresses, or address ranges that need to be able to access the service.

Beyond the security considerations, these alternative Kubernetes services implementations (e.g., eBPF-based dataplanes) provide other advantages over kube-proxy's implementation, such as:

- Improved performance when running with very high numbers of services, including reduced first packet latencies and reduced control plane CPU usage
- Direct server return (DSR), which reduces the number of network hops for return traffic

We will look at DSR more closely next, since it does have some security implications.

Direct Server Return

DSR allows the return traffic from the destination pod to flow directly back to the client rather than going via the original ingress node. There are several network plugins that are able to replace kube-proxy's service handling with their own implementations that support DSR. For example, a eBPF dataplane that includes native service handling and (optionally) can use DSR for return traffic is illustrated in Figure 9-7.

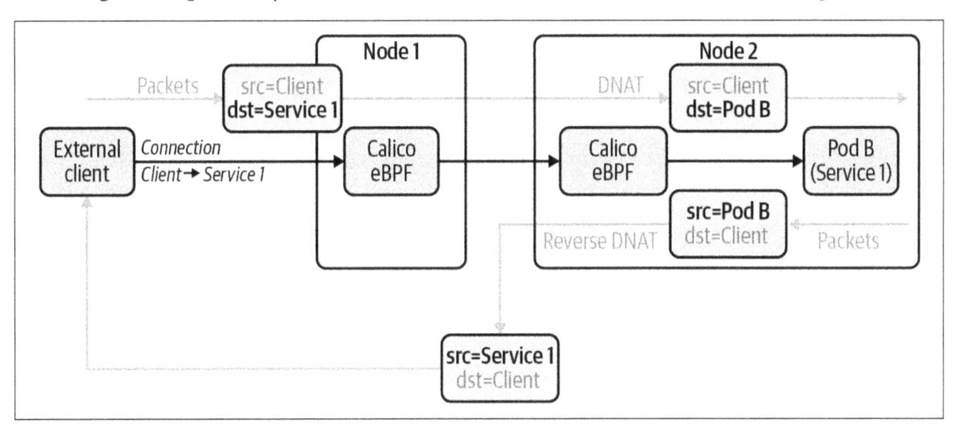

Figure 9-7. Network path for a Kubernetes service with direct server return

Eliminating one network hop for the return traffic reduces:

- The overall latency for the service (since every network hop introduces latency)
- The CPU load on the original ingress node (since it is no longer dealing with return traffic)
- The east-west network traffic within the cluster

For particularly network-intensive or latency-sensitive applications, this can be a big win. However, there are also security implications of DSR. In particular, the underlying network may need to be configured with fairly relaxed reverse path filtering (RPF) settings.

RPF is a network routing mechanism that blocks any traffic from a particular source IP address where there is not a corresponding route to that IP address over the same link. That is, if the router doesn't have a route that says it can send traffic to a particular IP address over the network link, then it will not allow traffic from that IP address over the network link. RPF makes it harder for attackers to "spoof" IP addresses—i.e., pretend to be a different IP address than what the device has been allocated.

In the context of DSR and Kubernetes services, Figure 9-7 illustrates a few key points:

- If the service is being accessed via a node port on Node 1, then the return traffic from Node 2 will have the source IP address of Node 1. So the underlying network must be configured with relaxed RPF settings, otherwise the network will filter out the return traffic because the network would not normally route traffic to Node 1 via the network link to Node 2.
- If the service is being accessed via service IP advertisement (e.g., sending traffic directly to a service's cluster IP, external IP, or load balancer IP), then the return traffic from Node 2 will have the source IP address of the service IP. In this case, no relaxation of RPF is required, since the service IP should be advertised from all nodes in the cluster, meaning the network will have routes to the service IP via all nodes. We'll cover service IP advertising in more detail later in this chapter.

As explained earlier, DSR is an excellent optimization that you can use, but you need to review your use case and ensure that you are comfortable with disabling the RPF check.

Limiting Service External IPs

So far in this chapter we have focused on how service types and implementations impact how network policy can be used to restrict access to services to only the specific client IP addresses or address ranges that need to be able to access each service.

Another important security consideration is the power associated with users who have permissions to create or configure Kubernetes services. In particular, any user who has RBAC permissions to modify a Kubernetes service effectively has control over which pods that service is load balanced to. If used maliciously, this could mean the user is able to divert traffic that was intended for a particular microservice to their own malicious pods.

As Kubernetes services are namespaces resources, this rarely equates to a genuine security issue for mainline service capabilities. For example, a user who has been granted permissions to define services in a particular namespace will typically also have permission to modify pods in that namespace. So for standard service capabilities such as handling of cluster IPs, node ports, or load balancers, the permissions to define and modify services in the namespace doesn't really represent any more trust than having permissions to define or modify pods in the namespace.

There is one notable exception, though, which is the ability to specify external IPs for services. The externalIP field in a service definition allows the user to associate an arbitrary IP address with the service. Connections to this IP address received on any node in the cluster are load balanced to the pods backing the service.

The normal use case is to provide an IP-oriented alternative to node ports that can be used by external clients to connect to a service. This use case usually requires special handling within the underlying network in order to route connections to the external IP to the nodes in the cluster. This may be achieved by programming static routes into the underlying network, or in a BGP-capable network, using BGP to dynamically advertise the external IP. (See the next section for more details on advertising service IP addresses.)

Like the mainline service capabilities, this use case is relatively benign in terms of the level of trust for users. It allows them to offer an additional way to reach the pods in the namespaces they have permission to manage, but does not interfere with traffic destined to pods in other namespaces.

However, just like with node ports, connections from pods to external IPs are also intercepted and load balanced to the service backing pods. As Kubernetes does not police or attempt to provide any level of validation on external IP addresses, this means a malicious user can effectively intercept traffic to any IP address, without any namespace or other scope restrictions. This is an extremely powerful tool for a malicious user and represents a correspondingly large security risk.

If you are following the best practice of having default deny–style policies for both ingress and egress traffic that apply across all pods in the cluster, then this significantly hampers the malicious user's attempt to get access to traffic that should have been between two other pods. However, although the network policy will stop them from accessing the traffic, it doesn't stop the service load balancing from diverting the traffic from its intended destination, which means that the malicious user can effectively block traffic between any two pods even though they cannot receive the traffic themselves.

So in addition to following network policy best practices, it is recommended to use an admission controller to restrict which users can specify or modify the externalIP field. For users who are allowed to specify external IP addresses, it may also be desirable to restrict the IP address values to a specific IP address range that is deemed safe (i.e., a range that is not being used for any other purpose). For more discussion of admission controllers, see Chapter 8.

Advertising Service IPs

One alternative to using node ports or network load balancers is to advertise service IP addresses over BGP. This requires the cluster to be running on an underlying network that supports BGP, which typically means an on-prem deployment with standard top-of-rack routers.

For example, Calico supports advertising the service clusterIP, loadBalancerIP, or externalIP for services configured with one. If you are not using Calico as your network plug-in, then MetalLB provides similar capabilities that work with a variety of different network plug-ins.

Advertising service IPs effectively allows the underlying network routers to act as load balancers, without the need for an actual network load balancer.

The security considerations for advertising service IPs are equivalent to those of normal node port– or load balancer–based services discussed earlier in this chapter. When using kube-proxy, the original client IP address is obscured by default, as illustrated in Figure 9-8.

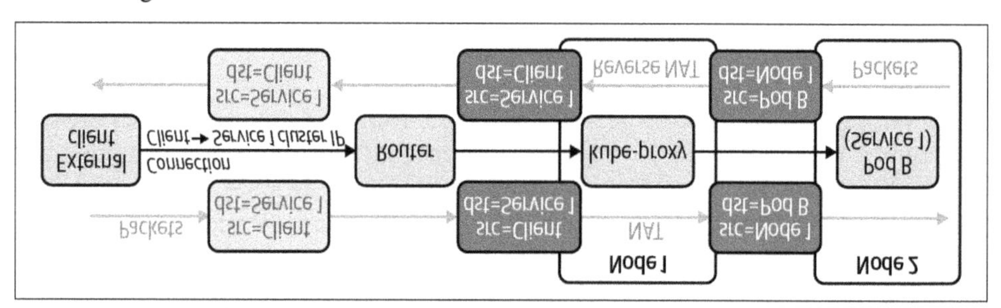

Figure 9-8. Network path for a Kubernetes service advertising Cluster IP via BGP

This behavior can be changed using `externalTrafficPolicy:local`, which (at the time of writing) is supported by kube-proxy for both loadBalancerIP and externalIP addresses but not clusterIP addresses. However, it should be noted that when using `externalTrafficPolicy:local`, the evenness of the load balancing becomes topology-dependent. To circumvent this, pod anti-affinity rules can be used to ensure even distribution of backing pods across your topology, but this does add some complexity to deploying the service.

Alternatively, a network plug-in with native service handling (replacing kube-proxy) that supports source IP address preservation can be used. This combination can be very appealing for on-prem deployments due to its operational simplicity and removal of the need to build network load balancer appliances into the network topology.

Understanding Kubernetes Ingress

Kubernetes Ingress builds on top of Kubernetes services to provide load balancing at the application layer, mapping HTTP and HTTPS requests with particular domains or URLs to Kubernetes services. Kubernetes Ingress can be a convenient way of exposing multiple microservices via a single external point of contact, if for example multiple microservices make up a single web application. In addition, they can be used to terminate SSL/TLS (for receiving HTTPS encrypted connections from external clients) before load balancing to the backing microservices.

The details of how Kubernetes Ingress is implemented depend on which Ingress Controller you are using. The Ingress Controller is responsible for monitoring Kubernetes Ingress resources and provisioning/configuring one or more ingress load balancers to implement the desired load balancing behavior.

Unlike Kubernetes services, which are handled at the network layer (L3–L4), ingress load balancers operate at the application layer (L5–L7). Incoming connections are terminated at the load balancer so it can inspect the individual HTTP/HTTPS requests. The requests are then forwarded via separate connections from the load balancer to the chosen service. As a result, network policy applied to the pods backing a service sees the ingress load balancer as the client source IP address, rather than the original external client IP address. This means they can restrict access to only allow connections from the load balancer, but cannot restrict access to specific external clients.

To restrict access to specific external clients, the access control needs to be enforced either within the application load balancer or in front of the application load balancer. In case you choose an IP-based access control, it needs to happen before the traffic is forwarded to the backing services. How you do this depends on the specific Ingress Controller you are using.

Broadly speaking, there are two types of ingress solutions:

In-cluster ingress
: Ingress load balancing is performed by pods within the cluster itself.

External ingress
: Ingress load balancing is implemented outside of the cluster by appliances or cloud provider capabilities.

Now that we have covered Kubernetes Ingress, let's review ingress solutions.

In-cluster ingress solutions

In-cluster ingress solutions use software application load balancers running in pods within the cluster itself. There are many different Ingress Controllers that follow this pattern. For example, the NGINX Ingress Controller instantiates and configures NGINX pods to act as application load balancers.

The advantages of in-cluster ingress solutions are that:

- You can horizontally scale your Ingress solution up to the limits of Kubernetes.
- There are many different in-cluster Ingress Controller solutions, so you can choose the Ingress Controller that best suits your specific needs— for example, with particular load balancing algorithms, security options, or observability capabilities.

To get your ingress traffic to the in-cluster Ingress pods, the Ingress pods are normally exposed externally as a Kubernetes service, as illustrated in Figure 9-9.

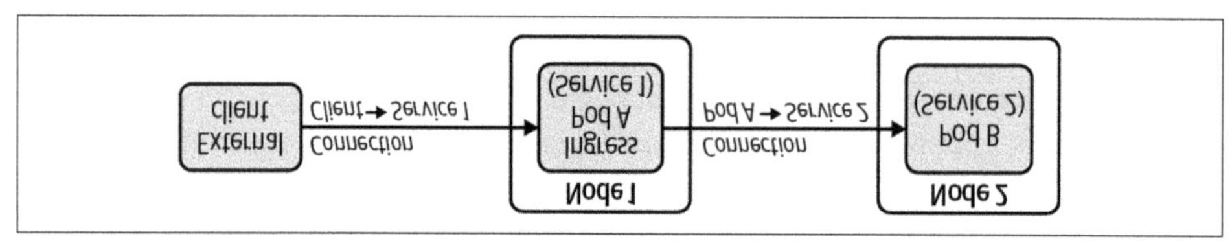

Figure 9-9. An example of an in-cluster ingress implementation in a Kubernetes cluster

This means you can use any of the standard ways of accessing the service from outside of the cluster. One common approach is to use an external network load balancer or service IP advertisement, along with one of the following:

- A network plug-in with native Kubernetes service handling that always preserves the original client source IP
- `externalTrafficPolicy:local` (and pod anti-affinity rules to ensure even load balancing across the ingress pods) to preserve the original client source IP

Network policy applied to the ingress pods can then restrict access to specific external clients as described earlier in this chapter, and the pods backing any microservices being exposed via Ingress can restrict connections to just those from the ingress pods.

External ingress solutions

External ingress solutions use application load balancers outside of the cluster, as illustrated in Figure 9-10.

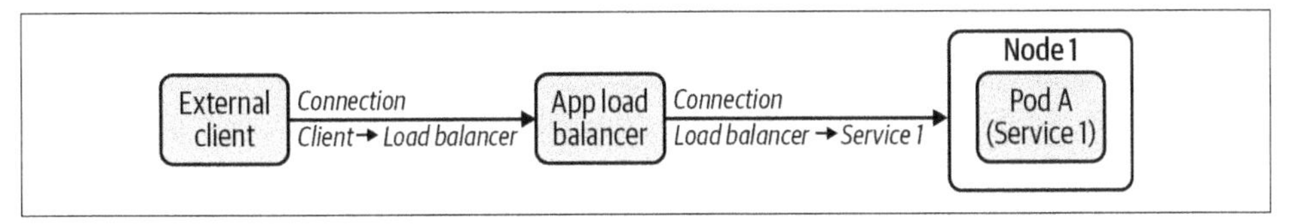

Figure 9-10. An example of an external ingress in a Kubernetes cluster

The exact details and features depend on which Ingress Controller you are using. Most public cloud providers have their own Ingress Controllers that automate the provisioning and management of the cloud provider's application load balancers to provide ingress.

Most application load balancers support a basic operation mode of forwarding traffic to the chosen service backing pods via the node port of the corresponding service. In addition to this basic approach of load balancing to service node ports, some cloud providers support a second mode of application load balancing that load-balances directly to the pods backing each service, without going via node ports or other kube-proxy service handling. This has the advantage of eliminating the potential second network hop associated with node ports load-balancing to a pod on a different node.

The main advantage of an external ingress solution is that the cloud provider handles the operational complexity of the ingress for you. The potential downsides are as follows:

- The set of features available is usually more limited, compared with the rich range of available in-cluster ingress solutions. For example, if you require a specific load balancing algorithm, security controls, or observability capabilities, these may or may not be supported by the cloud provider's implementation.
- The maximum supported number of services (and potentially the number of pods backing the services) is constrained by cloud provider–specific limits. For example, if you are operating at very high scales, with hundreds of pods backing a service, you may exceed the application layer load balancer's maximum limit of IPs it can load balance to in this mode. In this case switching to an in-cluster ingress solution is likely the better fit for you.

- Since the application load balancer is not hosted within the Kubernetes cluster, if you need to restrict access to specific external clients, you cannot use Kubernetes network policy and instead must use the cloud provider's specific mechanisms. It is still possible to follow the best practices laid out at the start of this chapter, but doing so will be cloud provider–specific and will likely introduce a little additional operational complexity, compared with being able to use native Kubernetes capabilities independent of the cloud provider's capabilities and APIs.

In this section we covered how Kubernetes ingress works and the solutions available. We recommend you review the sections and decide if the in-cluster ingress solution works for you or if you should go with an external ingress solution.

Conclusion

In this chapter, we covered the topic of exposing Kubernetes services outside the cluster. The following are the key concepts covered:

- Kubernetes services concepts like direct pod connections, advertising service IPs, and node ports are techniques you can leverage to expose Kubernetes services outside the cluster.
- We recommend using an eBPF-based dataplane to optimize the ingress path to route traffic to the pods hosting the service backend.
- An eBPF dataplane is an excellent alternative to the default Kubernetes services implementation, kube-proxy, due to its ability to preserve source IP to the pod.
- The choice of a Kubernetes ingress implementation will depend on your use case. We recommend that you consider an in-cluster ingress solution as it is more native to Kubernetes and will give you more control than using an external ingress solution.

We hope you are able to leverage these concepts based on your use case as you implement Kubernetes services.

CHAPTER 10

Encryption of Data in Transit

As you move mission-critical workloads to production, it is very likely that you will need to encrypt data in transit. It is a very important requirement for certain types of data to meet compliance requirements and also a good security practice.

Encryption of data in transit is a requirement defined by many compliance standards, such as HIPAA, GDPR, and PCI. The specific requirements vary somewhat; for example, PCI DSS (Payment Card Industry Data Security Standard) has rules around encryption of cardholder data while in transit. Depending on the specific compliance standard, you may need to ensure data in transit between the applications or microservices hosted in Kubernetes is encrypted using a recognized strong encryption algorithm.

And depending on the architecture of your application or microservices, it may be that not all data being sent over the network is classified as sensitive, so theoretically you might strictly only need to encrypt a subset of the data in transit. However, from the perspective of operational simplicity and ease of compliance auditing, it often makes sense to encrypt all data in transit between your microservices, rather than trying to do it selectively.

Even if you do not have strong requirements imposed by external compliance standards, it can still be a very good practice to encrypt data in transit. Without encryption, malicious actors with network access could see sensitive information. How you assess this risk may vary depending on whether you are using public cloud or on-prem/private cloud infrastructure, and the internal security processes you have in place as an organization. In most cases, if you are handling sensitive data, then you should really be encrypting data in transit.

If you are providing services that are accessed by clients on the public internet, then the standard practice of using HTTPS applies to Kubernetes. Depending on your microservice architecture, these HTTPS connections can be terminated on the destination microservice, or they may be terminated by a Kubernetes Ingress solution, either as in-cluster Ingress pods (e.g., when using the NGINX Ingress Controller) or out-of-cluster application load balancers (e.g., when using the AWS Load Balancer Controller). Note that if using an out-of-cluster application load balancer, it's important to still make sure that the connection from the load balancer to the destination microservice uses HTTPS to avoid an unencrypted network hop.

Within the cluster itself, there are three broad approaches to encrypting data in transit:

- Build encryption capabilities into your application/microservices code.
- Use sidecar- or service mesh–based encryption to encrypt at the application layer without needing code changes to your applications/microservices.
- Use network-level encryption, again without the need for code changes to your applications/microservices.

We will now explore the pros and cons of each approach.

Building Encryption into Your Code

There are libraries to encrypt network connections for most programming languages, so in theory you could choose to build encryption into your microservices as you build them. For example, you could use HTTPS SSL/TLS or even mTLS (mutual TLS) to validate the identity of both ends of the connection.

However, this approach has a number of drawbacks:

- In many organizations, different microservices are built using different programming languages, with each microservice development team using the language that is most suited for that particular microservice and team's expertise. For example, a frontend web UI microservice might be written using Node.js, and a middle-layer microservice might be written in Python or Golang. As each programming language has its own set of libraries available for encryption, this means that the implementation effort increases, potentially with each microservices team having to implement encryption for their microservice rather than being able to leverage a single shared implementation across all microservices.
- Building on this idea of not having a single shared implementation for encryption, the same applies to configuration of the microservices, in particular, how the microservice reads its credentials required for encryption.

- In addition to the effort involved in developing and maintaining all this code, the more implementations you have, the more likely it is that one of the implementations will have bugs in it that lead to security flaws.
- It is not uncommon for older versions of encryption libraries to have known vulnerabilities that are fixed in new versions. By the time a new version is released to address any newly discovered vulnerability, the vulnerability is public knowledge. This in turn increases the number of attacks targeted at exploiting the vulnerability. To mitigate against this, it is essential to update any microservices that use the library as soon as possible. If you are running many microservices, this may represent a significant development and test effort, since the code for each microservice needs to be updated and tested individually. On top of that, if you don't have a lot of automation built into your CI/CD process, then there may also be the operational headache of updating each microservice version with the live cluster.
- Many microservices are based on third-party open source code (either in part or for the whole of the microservice). Often this means you are limited to the specific encryption options supported by the third-party code, and in many cases the specific configuration mechanisms that the third-party code supports. You also become dependent on the upstream maintainers of the third-party code to keep the open source project up to date and address vulnerabilities as they are discovered.
- Finally, it is important to note that there is often operational overhead when it comes to provisioning encryption settings and credentials across disparate implementations and their various configuration paradigms.

The bottom line, then, is that while it is possible to build encryption into each of your microservices, the effort involved and the risk of unknowingly introducing security flaws (due to code or design issues or outdated encryption libraries) can make this approach feel pretty daunting and unattractive.

Sidecar or Service Mesh Encryption

An alternative architectural approach to encrypting traffic between microservices at the application layer is to use the sidecar design pattern. The sidecar is a container that can be included in every Kubernetes pod alongside the main container(s) that implement the microservice. The sidecar intercepts connections being made to/from the microservice and performs the encryption on behalf of the microservice, without any code changes in the microservice itself. The sidecar can either be explicitly included in the pod specification or it can be injected into the pod specification using an admission controller at creation time.

Compared to building encryption into each microservice, the sidecar approach has the advantage that a single implementation of encryption can be used across all microservices, independent of the programming language the microservice might have been written in. It means there is a single implementation to keep up to date, which in turn makes it easier to roll out vulnerability fixes or security improvements across all microservices with minimal effort.

You could in theory develop such a sidecar yourself. But unless you have some niche requirement, it would usually be better to use one of the many existing free open source implementations already available, which have had a significant amount of security review and in-field hardening.

One popular example is the Envoy proxy, originally developed by the team at Lyft, which is often used to encrypt microservice traffic using mTLS (mutual TLS). Mutual TLS means that both the source and destination microservices provide credentials as part of setting up the connection, so each microservice can be sure it is talking to the other intended microservice. Envoy has a rich configuration model, but does not itself provide a control or management plane, so you would need to write your own automation processes to configure Envoy to work in the way you want it to.

Rather than writing this automation yourself, an alternative approach is to use one of the many service mesh solutions that follow a sidecar model. For example, the Istio service mesh provides a packaged solution using Envoy as the sidecar integrated with the Istio control and management plane. Service meshes provide many features beyond encryption, including service routing and visibility. While service meshes are becoming increasingly popular, a widely acknowledged potential downside of their richer feature set is it can introduce operational complexity, or make the service mesh harder to understand at a nuts-and-bolts level with a greater number of moving parts. Another downside is the security risk associated with the sidecar design pattern where the sidecar is part of every application pod and there is additional complexity of managing sidecars (for example, a CVE may require you to update sidecars, and this is not a trivial update as it impacts all applications).

Network-Layer Encryption

Implementing encryption with the microservice or using a sidecar model is often referred to as *application-layer encryption*. Essentially, the application (microservice or sidecar) handles all of the encryption, and the network is just responsible for sending and receiving packets, without being aware the encryption is happening at all.

An alternative to application-layer encryption is to implement encryption within the network layer. From the application's perspective, it is sending unencrypted data, and it is the network layer that takes responsibility for encrypting the packets before they are transmitted across the network.

One of the main standards for network-layer encryption that has been widely used throughout the industry for many years is IPsec. Most IPsec implementations support a broad range of encryption algorithms, such as AES encryption, with varying key lengths. IPsec is often paired with IKE (Internet Key Exchange) as a mechanism for managing and communicating the host credentials (certificates and keys) that IPsec needs to work. There are a number of open source projects, such as the popular strongSwan solution, that provide IKE implementations and make creating and managing IPsec networks easier.

Some enterprises choose to use solutions such as strongSwan as their preferred solution for managing IPsec, which they then run Kubernetes on top of. In this case Kubernetes is not really aware of IPsec. Even with projects such as strongSwan helping to make IPsec easier to set up and manage, many regard IPsec as being quite heavyweight and tricky to manage from an overall operational perspective.

One alternative to IPsec is WireGuard. WireGuard is a newer encryption implementation designed to be extremely simple yet fast, using state-of-the-art cryptography. Architecturally, it is simpler, and initial testing indicates that it does outperform IPsec in various circumstances. It should be noted though that development continues on both WireGuard and IPsec, and in particular as advances are made to cryptographic algorithms, the comparative performance of both will likely evolve.

Rather than setting up and managing IPsec or WireGuard yourself, an operationally easier approach for most organizations is to use a Kubernetes network plug-in with built-in support for encryption. There are a variety of Kubernetes network plug-ins that support different types of encryption, with varying performance characteristics.

If you are running network-intensive workloads, then it is important to consider the performance cost of encryption. This cost applies whether you are encrypting at the application layer or at the network layer, but the choice of encryption technology can make a significant difference to performance. For example, Figure 10-1 shows independent benchmark results for four popular Kubernetes network plug-ins (the most recent benchmarks available at the time of writing, published in 2020).

Using a Kubernetes network plug-in that supports encryption is typically significantly simpler from an operational standpoint, with many fewer moving parts than adopting a service mesh, and significantly less effort than building encryption into your application/microservice code. If your primary motivation for adopting a service mesh is security through encryption, then using a Kubernetes network plug-in that supports network-layer encryption along with Kubernetes network policies is likely to be significantly easier to manage and maintain. Please note that we cover other aspects of service mesh like observability in Chapter 5.

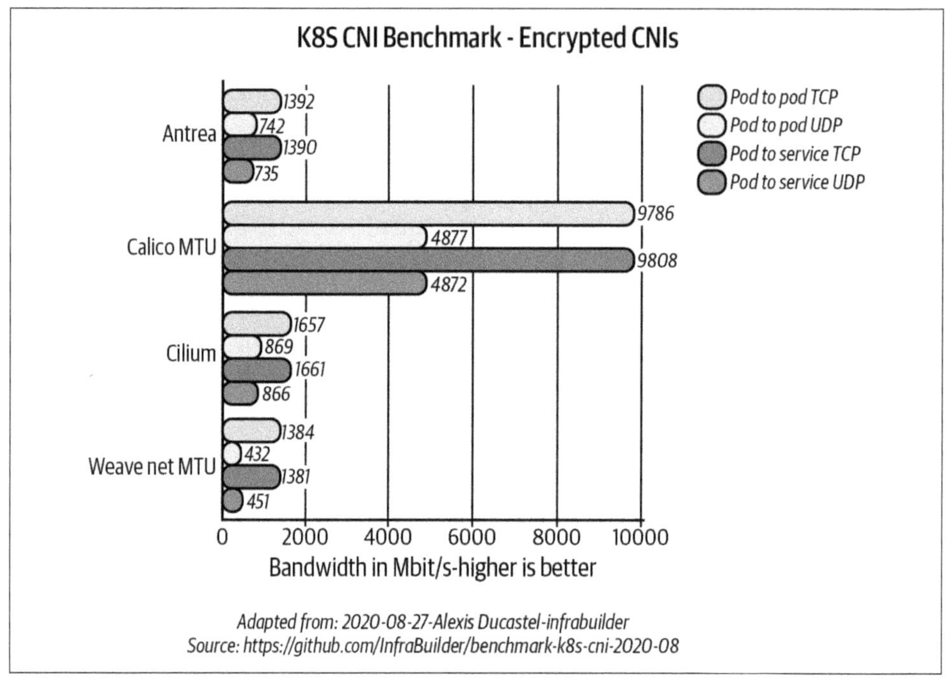

Figure 10-1. Benchmark results for encryption in Kubernetes

Conclusion

In this chapter we presented various options to implement encryption of data in transit and various approaches to implement encryption in a Kubernetes cluster. We hope this enables you to pick the option most suited for your use case. Here are a few things to remember:

- As you move mission-critical workloads to production, for certain types of data, you will need to implement encryption for data in transit. We recommend implementing encryption of data in transit even if compliance requirements do not require you to encrypt all data.
- We covered the well-known methods of how you can implement encryption: application-layer encryption, sidecar-based encryption using a service mesh, and network-layer encryption.
- Based on operational simplicity and better performance, we recommend network-layer encryption.

CHAPTER 11

Threat Defense and Intrusion Detection

In this chapter we will explore how you can implement threat defense for your Kubernetes cluster. We have covered the stages of your Kubernetes deployment (build, deploy, runtime) in earlier chapters. This chapter focuses on threat defense, which is security for the runtime stage. We will cover the following concepts to help you understand threat defense in a Kubernetes cluster and why you need it.

- Threat defense for a Kubernetes cluster, including why you need it and how it differs from traditional security
- Intrusion detection for Kubernetes
- Advanced threat defense techniques

Let's explore each of these in detail. We start with threat defense and why it is important.

Threat Defense for Kubernetes (Stages of an Attack)

To understand threat defense, a great place to start is to review the cybersecurity kill chain, which breaks down an attack into several stages. This is then used to build a strategy to defend against the attack. The cyber kill chain has the following stages:

Reconnaissance
: Adversaries probe the target and gather information.

Weaponization
: The adversary creates a method to attack, which could be a new vulnerability, a variant of an existing vulnerability, or a simple exploit of an insecure configuration.

Delivery
: The adversary creates a method to deliver the vulnerability or exploit to the target or a location that can be used to attack the target.

Exploitation
: The adversary implements methods to trigger the attack.

Installation
: The adversary installs the malware and typically software to create a backdoor to communicate with the malware.

Command and Control
: The adversary establishes a communication channel with the malicious software to control the software.

Actions on Objective
: The adversary achieves the intended outcome of the attack (e.g., stealing data, encryption of data, etc.).

Several organizations have adapted this framework to incorporate real-world attacks and more use cases. Microsoft has adapted it for Kubernetes as described in its blog post "blog"Secure Containerized Environments with Updated Threat Matrix for Kubernetes" (*https://oreil.ly/gebGs*). Let's review the kill-chain stages (threat matrix) that are specific to Kubernetes:

Initial access
: The adversary uses various exposed interfaces in your Kubernetes deployments (for example, Kubeflow) via stolen credentials, compromised images, or other application vulnerabilities.

Execution
: The adversary executes a malicious command or software in your cluster. There are several ways this can happen; some known techniques are creating a new container in your cluster, running an additional container in any pod as a sidecar, and using a known application vulnerability to execute malicious commands.

Persistence
: The adversary will try to persist the malicious software in the Kubernetes cluster so it can be accessed at a later time. This is typically achieved by creating a writage storage path on the host, leveraging Kubernetes scheduled jobs (known as cron jobs) to periodically run malicious software, or in some cases, compromising the Kubernetes admission controller so requests to the API server can be tampered with to carry out an attack. In this stage adversaries will also try to establish a backdoor communication channel to their control server so they can control the malicious software. This is known as a command-and-control server (C&C server).

Privilege escalation
: The adversary gains privileged access by leveraging a resource in your cluster that has privileged access. For example, they might run malicious software in a container with privileged access by exploiting a vulnerability in the privileged container.

Defense evasion
: The adversary works to keep the attack undetected by using techniques like clearing logs or deleting events, so detection systems using logs and events do not detect the presence of malicious activity. Another technique that is used is to exploit a vulnerability in only one pod of a Kubernetes deployment backed by many pods, and use that to further the attack.

Credential access
: The adversary works to get access to credentials (Kubernetes secrets) in your cluster. In case you are using managed services, the cloud provider offers a token to access cloud resources, and this token is accessible to certain privileged pods and service accounts. The adversary will use impersonation or privilege escalation to gain access to the credentials and then use access to cloud resources to further the attack.

Discovery
: The adversary will work to do a reconnaissance of the cluster network to understand what is running in your cluster. This can be achieved by using tools for network mapping like Nmap (*https://nmap.org*) on Linux systems, or access to the Kubenetes dashboard. This stage is the precursor to an important stage of the attack where adversaries can move around in your cluster to find what they are seeking.

Lateral movement
: By the time the attack reaches this stage, the attack is fairly advanced, and the adversary has an established presence; they now will use the installed malicious software to access other pods and resources in the network. A couple of common techniques are to spoof IP addresses or domain names of other pods and impersonate other pods to get past segmentation rules inside the cluster. The adversaries also look for other applications running inside the cluster, as they now have access to them. During this stage the malicious software is communicating with the command-and-control server to get instructions to further the attack. In this stage adversaries rely on overloading well-known protocols like DNS or HTTP to send command-and-control requests as a part of these protocols, which allows them to bypass perimeter security–based controls as the traffic looks like a normal DNS or HTTP request.

Impact
: This is the final stage of the attack, where the outcome is usually the stealing of sensitive data. This is achieved by a technique called data exfiltration, encryption of data for ransomware, or even using resources for cryptomining.

Threat defense comprises a set of techniques that help you defend against each of these stages and enable you to defend against attacks. It can be overwhelming to think about all these stages and techniques adversaries can use. We want to mention that while adversaries need to succeed in most (if not all) of these stages to carry out a successful attack, you only need to block them in any one stage to thwart the attack. So the odds are in your favor. Understanding these stages and how they apply to Kubernetes is a first step in building an effective defense mechanism. Adversaries are always innovating, and therefore you should focus on all the stages and use tools and techniques relevant to each stage to give yourself the greatest chance to successfully thwart attacks.

In Chapter 2, we covered infrastructure security, showing you how to create secure infrastructure for running your workloads. In Chapter 3, we covered best practices and techniques you can use to securely deploy workloads. Chapter 4 covers security policies you can apply to your workloads to secure the workload runtime environment, and Chapter 6 covers how you can apply network policy to implement network access control for your workloads. We recommend you review these chapters in the context of the kill-chain stages described here. You will find that these techniques are very effective with the initial access, privilege escalation, credential access, persistence, and execution stages.

We will now describe tools and techniques you can use to secure the other stages of the kill chain. It is important to note that Kubernetes is a distributed system and its cluster network is crucial to its operation; therefore, securing the network is a very effective technique. For example, a successful privilege escalation or a successful exploit of an application vulnerability is rendered ineffective if the adversary cannot further the attack due to an inability to use the cluster network for discovery, command and control, lateral movement, or data exfiltration. It is not enough to have network segmentation based on IP addresses/ports, as adversaries will find ways to further an attack even with techniques like network segmentation protecting your cluster. For example, you need to allow HTTP traffic to your service and pods backing the service, so the attack can be a part of the HTTP header that triggers a privilege escalation.

Now that we have covered concepts for threat defense, let's explore intrusion detection.

Intrusion Detection

In this section we will cover intrusion detection and how it applies to Kubernetes clusters. To understand this, we will review the various methods of intrusion and the role of an intrusion detection system.

Intrusion Detection Systems

An intrusion detection system (IDS), as the name suggests, is a system that monitors network activity, detects anomalous patterns, and reports suspicious behavior. These systems also monitor violations to existing controls (like network policy, host hardening) and report these violations. The response actions for an IDS are the following:

Alerting
: Generate an alert and send it to a SIEM for further analysis and action.

Intrusion prevention
: The system takes action to prevent the intrusion by leveraging existing controls (e.g., network policies, Kubernetes pod security policies, host hardening policies) and redirecting the attack to canary resources especially set up to analyze these types of attack. When an IDS is also able to prevent the intrusion, it is called an intrusion prevention system (IPS).

A good intrusion detection system should be able to associate a set of related anomalies by tracking the behavior of a system. We recommend you review user and entity behavior analytics (UEBA) and how it applies to security. Microsoft Azure UEBA (*https://oreil.ly/LqYWk*) is an excellent resource for you to review. Please note that for Kubernetes, entity behavioral analytics is applicable. The details of how to implement it are outside the scope of this book, but UEBA helps in reducing the number of alerts and generating high-fidelity alerts. Please note that more alerts is not necessarily good; they cause downstream systems (e.g., SIEMs) to be immune to alerts. Later we will review how to leverage machine learning systems to generate high-fidelity alerts.

We will now review intrusion detection methods and how they apply to Kubernetes clusters.

IP Address and Domain Name Threat Feeds

As explained in the cybersecurity kill chain, adversaries will often use malicious software to contact a server that is controlled by them. These servers are used to remotely control the malicious software, get information about the system, download more software, and further the attack. Security research teams around the world review attacks and identify known C&C servers by IP address/domains. These are published as threat feeds and as a part of the indicators of compromise and are regularly updated. There are several well-known threat feeds both open source and commercial.

STIX (*https://oreil.ly/6mw8s*) is a well-known standard to describe threat intelligence, and TAXII (*https://oreil.ly/w9DSn*) is the standard to deliver the intelligence. You can find several open source engines that parse the STIX and TAXII feeds and provide intelligence (e.g., AlienVault). Feodo tracker (*https://oreil.ly/cOccW*) and Snort (*https://oreil.ly/1isJX*) are examples of open source feeds that provide IP address block lists.

Sometimes adversaries will use VPNs (virtual private networks), which are overlay networks that run over physical networks and are useful to conceal a user's location. Tor is another well-known overlay network that is used for this purpose. Similar to threat feeds for C&C servers, feeds are available for known VPNs (*https://oreil.ly/RGtxT*) and IPs from the Tor network (*https://oreil.ly/VHGkZ*).

We will now cover how you can use these feeds in your Kubernetes cluster to implement IDS/IPS. Figure 11-1 shows a sample implementation of applying suspicious IP addresses and domains in your Kubernetes cluster.

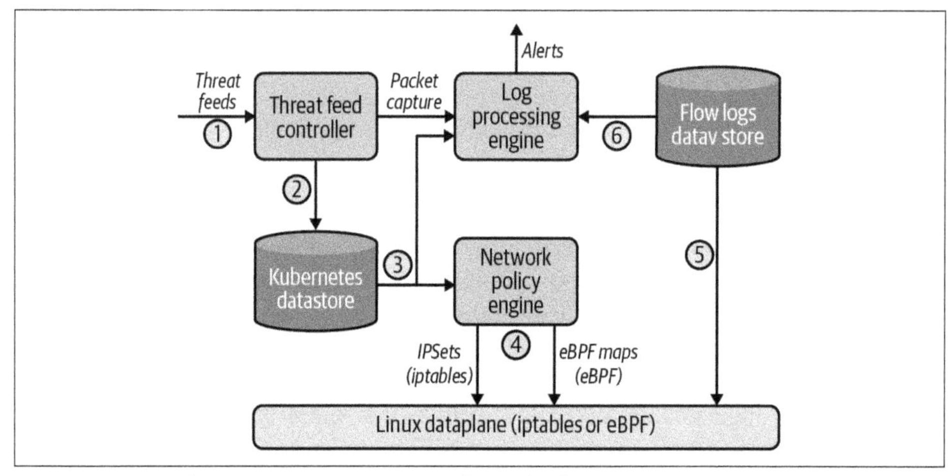

Figure 11-1. Implementing IDS/IPS using threat feeds

Figure 11-1 shows what you need to implement support for threat feeds in your cluster and also describes the high-level workflow to achieve IDS/IPS capabilities. The figure shows the following components as a part of your Kubernetes cluster.

Threat feed controller

This component is responsible for retrieving threat feeds from a configured source (typically a URL). This can be implemented in many ways. For the purposes of this discussion, we assume that is a pod that watches a configuration resource (example in a moment) and reaches out to the specified URL and stores the threat feed data in the Kubernetes datastore for other components. The following is an example of configuration for the threat feed controller:

```
apiVersion: projectcalico.org/v3
kind: GlobalThreatFeed
metadata:
  name: sample-global-threat-feed
spec:
  content: IPSet
  pull:
    http:
      url: https://an.example.threat.feed/blacklist
  globalNetworkSet:
    labels:
      security-action: block
```

In this example, the threat feed controller is configured to pull the threat feed from the specified URL and then stores the list of IP addresses as a custom Kubernetes resource named globalnetworkset. This is then used by the network policy implementation to enforce policies based on this resource. The following is an example of a policy that can be defined using the globalnetworkset resource:

```
apiVersion: projectcalico.org/v3
kind: GlobalNetworkPolicy
metadata:
  name: default.blockthreats
spec:
  tier: default
  selector: all()
  types:
  - Ingress
  ingress:
  - action: Deny
    source:
      selector: security-action == 'block'
```

Network policy engine

This is the component that implements network policies in your cluster and is used to define policies to block traffic to and from IP addresses that are a part of threat feeds. Please note that an IP address list can contain a large number of IP addresses and can change periodically. Therefore, the network policy implementation you choose should scale based on this requirement. As a hint, we recommend you pick an engine that supports the ipsets extension in an iptables-based dataplane, where matching sets of IP addresses is optimized. If you are using an eBPF-based dataplane, please ensure the implementation has support for eBPF maps to implement functionality equivalent to ipsets.

Log processing engine

This component is responsible for reporting flow logs from your cluster that contain IP addresses that match any IP addresses that are part of the threat feed and generate an alert. Please note this can be a resource-intensive operation, given the large amount of flow data. One way to address this is to have the network policy engine add an annotation to the flow log to name the feed that contained the IP address in the flow log when the dataplane detects a match. It is very efficient to do this operation inline instead of doing a match after data is collected.

Now that we understand the various components of the IDS, let's review the step-by-step operation:

1. The threat feed controller polls the threat feed periodically.
2. The threat feed controller processes the feed and creates the globalnetworkset resource in the Kubernetes datastore for the threat feed.
3. The network policy engine and the log processing engine read the threat feed.
4. If a network policy is defined, the network policy engine implements the network policy for the threat feed.
5. The log engine processes flow logs from the flow log datastore and generates alerts for flows matching IP addresses in the threat feed.

In step 4, we are able to prevent an intrusion, and step 5 is where we can detect an intrusion.

Special Considerations for Domain Name Feeds

As mentioned before, threat feeds can be a list of IP addresses or domain names. In case of domain names, the network policy engine must support domain name–based policies, and the log processing engine must support capturing domain names in flow logs and matching domain names from feeds in a flow log.

Note that the technique described detects and enables controls to protect against malicious activity, so it makes the implementation an intrusion prevention system.

Deep packet inspection

Deep packet inspection (DPI) is an intrusion detection technique where network traffic is inspected and matched against known malicious network traffic patterns. This requires inspection of the packet beyond Layer 3/Layer 4 and understanding of application protocols (e.g., HTTP, MySQL, etc.). Similar to threat feeds for IP and domains, signatures-based feeds are also available. OWASP Top 10 (*https://oreil.ly/270Fw*) and SANs Top 25 (*https://oreil.ly/OmO85*) are signatures for well-known application software risks and software vulnerabilities that can be used to detect malicious traffic in network and application layers of the packet. When you think about implementing DPI in your cluster, you need to consider a few factors.

First, where should you implement DPI? One option is to implement it at the ingress, which is the point where traffic enters/exits the cluster. Ingress is a resource in Kubernetes that allows users to expose services to clients outside the cluster. This topic is covered in depth in Chapter 8. For this discussion we assume that the services are exposed as URLs. So in this case you would implement DPI at the ingress (e.g., load balancers or on nodes for traffic going through node ports). This is a good option, but in this scenario the limitation is that you can detect a malicious network flow, but the information is not complete, as the detection is early in the cycle. You will not have visibility into which pod the malicious flow was destined to, and will have to review all possible destination pods and then co-relate the activity of each pod backing the service to understand the attack. Also, if the attack uses another mechanism to trigger the exploit (e.g., an API server, a kube-proxy vulnerability, or a node OS image vulnerability), the malicious flow originates inside the cluster and will not be detected at the ingress. Therefore, it is better to implement DPI for the service inside the cluster.

So if you choose the option to implement DPI inside the cluster for each service, you need to consider that the amount of traffic inside the cluster is very large due to the distributed nature of Kubernetes. This presents a challenge as DPI involves packet parsing and can potentially impact latency for applications as well as increase the resource utilization. In order to address the application latency challenge, we recommend you consider DPI as a mechanism to detect malicious activity and not prevent the attack. This means the DPI engine does not need to be inline and can work with a copy of each packet in the flow; the original packet flow is not impacted by this, and hence there is no impact to latency-sensitive applications. There is still the concern about resource utilization due to packet parsing. In order to address this, we recommend you use context in the Kubernetes cluster to select traffic for services that need DPI. This could be as simple as labeling services that are critical and have compliance requirements and enabling DPI using label-based selectors, or you can use DPI as a response action to anomalous traffic that is identified by a SIEM or your logging and alerting engine.

Another important consideration is that the DPI engine needs traffic to be unencrypted, so if you are using encryption for traffic inside the cluster (e.g., HTTPS), you need to implement decryption along with your DPI engine or choose an encryption technology like WireGuard, where you can implement DPI prior to encryption for egress traffic and after decryption for ingress traffic. You should consider a proxy like Envoy that allows traffic to be redirected to it, decrypted, inspected, encrypted, and sent to its destination.

Now that we have established that DPI needs to be implemented inside the cluster as an IDS mechanism and needs to be selectively enabled to limit resource consumption, let's explore a sample reference architecture for a DPI implementation for your Kubernetes cluster. Figure 11-2 shows a reference implementation. Before we review the reference implementation, we want to introduce some well-known IDS engines that are available for you to integrate in your cluster. Snort (*https://oreil.ly/yDteh*) and Suricata (*https://oreil.ly/1Ka1q*) are a couple of open source IDS engines that are available; however, you can choose any IDS engine that is suitable for your use case, including implementing your own IDS engine.

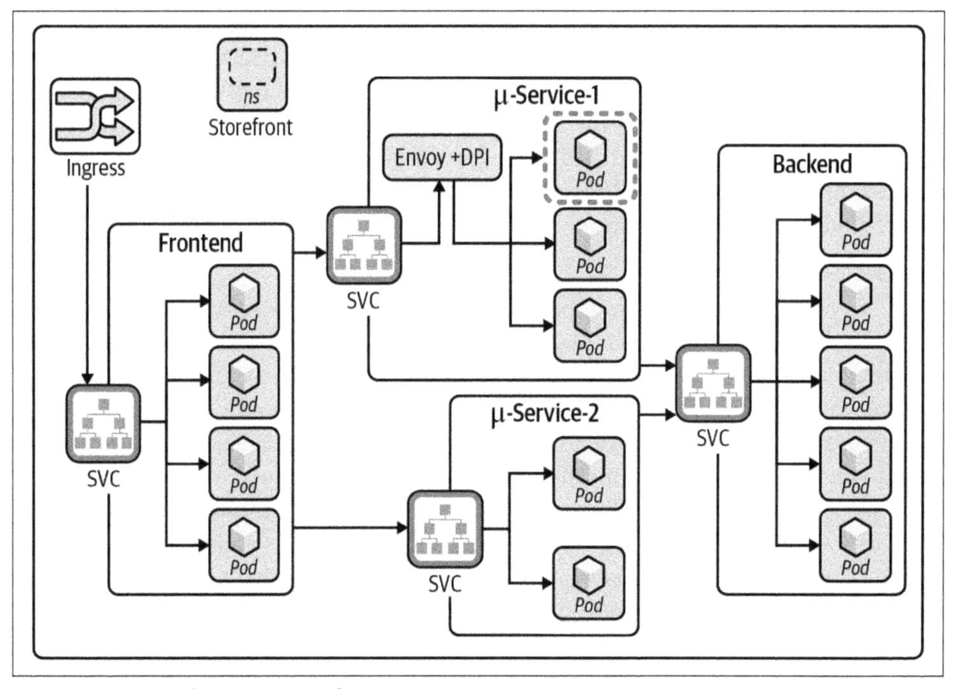

Figure 11-2. Implementation of DPI using Envoy

Figure 11-2 shows a Kubernetes cluster namespace that has a few microservices that are part of an application. The figure shows Envoy is deployed as a daemonset on every node. Envoy is a well-known proxy that is used in Kubernetes clusters to proxy traffic for analysis and for additional controls. We recommend you use Envoy as a transparent proxy, where it terminates connections destined to a pod backing a service and after analysis sends the traffic to the pod backing a service. In this case for the second microservice, the DPI engine is implemented as an integration to Envoy, and Envoy is configured to redirect traffic destined for pods backing the service to itself. DPI is then performed, and Envoy takes care of completing the connection. The transparent mode for Envoy means that the application sees no difference in the packet (e.g., TCP/IP header). As discussed before, if a malicious flow is detected by the DPI engine, the resulting alert will show the pod that received the flow, and it is then trivial to examine activity by that pod and co-relate it to the malicious flow. In Figure 11-2, DPI is enabled for traffic destined to one service as an example. DPI can be enabled for any service or a combination of pods; the approach is similar, but the difference is in how Envoy is configured for traffic redirection.

We recommend that you use Envoy with your DPI engine, but there are other ways to integrate a DPI engine in your Kubernetes cluster. For example, if you have a cluster running an eBPF dataplane, you can get copies of the packet in the BPF program and send them to the IDS engine for analysis. Likewise, if you are using VPP (*https://fd.io*) as the dataplane, it is also possible to integrate a DPI engine to inspect traffic.

You can choose the option that works for you, but it is important that you consider integration of a DPI engine in your Kubernetes cluster for signature-based malware detection. Next, let's examine logging and visibility and its role in the threat defense strategy.

Logging and visibility

A very important part of security is visibility of the activity in your cluster (e.g., pod creation, Kubernetes resource access/changes, application activity, network activity). This is achieved by enabling logging in your cluster. We cover log collection and metrics collection in detail in Chapter 5. For this section we want to reiterate the following as key aspects of logging and visibility:

- Traditional logging for network flows with the five-tuple is insufficient. You need to use a tool that supports Kubernetes context-rich logging where network flows between pods, deployments, replica sets, and services are part of the log collected.
- Logs at collection time need to be annotated with Kubernetes metadata like labels, policies in use, node information, and even process information (processes running in the container). This is important due to the ephemeral nature of Kubernetes; all of these change, and it is difficult to associate a malicious network flow with Kubernetes metadata when the network flows and the Kubernetes metadata are collected independent of each other.
- DNS activity logs are critical and must also be annotated with Kubernetes metadata as described earlier for network flow logs.
- Application protocol–based flow logs (e.g., HTTP header, MySQL) are also critical and again must be collected with Kubernetes metadata.
- Finally, Kubernetes audit logs (activity logs) are very important and must be collected, as these will help detect abnormal activity by malicious users (e.g., repeated denied access to a resource, creation of a service account, etc.).

There are several tools and mechanisms available to you to implement log collection. The cloud providers have logging capabilities (e.g., Stackdriver in Google, CloudWatch in AWS); you can also choose to implement logging using tools like Sysdig, Datadog, and Calico Enterprise, which offer logging capabilities with Kubernetes context. In addition to the log collection described earlier, the tool you choose must support the following simple capabilities that are critical to your IDS strategy:

- The tool must support an alerting capability that allows you to query logs and set up alerts for rule-based anomalies (e.g., excessive NXDOMAIN requests, imbalance in network throughput for a given protocol like HTTP or DNS between inbound and outbound traffic, unexpected connections to certain pods from certain namespaces, excessive network policy denied logs).
- The tools must support the baselining of various metrics (e.g., number of connections to a service, HTTP requests from a rare user-agent in the header) using basic machine learning techniques and report anomalies as alerts.
- The tools must support forwarding logs and alerts to an external SIEM (like Splunk, QRadar, Sumo Logic) or the cloud provider's security center (Azure Security Center).

The previous sections covered the collection and analysis of logs. While logging is available in most Kubernetes environments, you need to ensure that the tool you choose to implement logging is effective for your IDS strategy.

Advanced Threat Defense Techniques

In this section we will cover some advanced threat defense techniques you can use in your cluster. These are techniques that are designed to be effective in a Kubernetes environment, especially for detecting the lateral movement and exfiltration stages of the attack life cycle.

Canary Pods/Resources

The use of honeypots is a well-known technique to detect bad actors within your cluster and gain insight on what they are doing by exposing simulated or intentionally vulnerable applications in your cluster and monitoring access to these applications. These applications act as a canary to notify the blue team of the intrusion and stall the attacker's progress from reaching actual sensitive applications and data. Once the blue team is aware of the situation, the attack can be traced back to the initial vector. The attack can then be contained and even removed from the cluster.

Applying this technique in a Kubernetes environment works exceedingly well due to the declarative nature of applying manifests to deploy workloads. Whether the cluster is standalone or part of a complex pipeline, workload communications are defined by the application's code. Any communication that's not defined can be deemed suspicious at a minimum, and the source resource may have been compromised. By introducing fake workloads and services around production workloads, when a workload gets compromised, the attacker cannot differentiate between other real and fake workloads. The asymmetric knowledge between the attacker and the cluster operator makes it easy to detect lateral movement from compromised workloads.

Figure 11-3 shows an example of how this is achieved in a Kubernetes cluster.

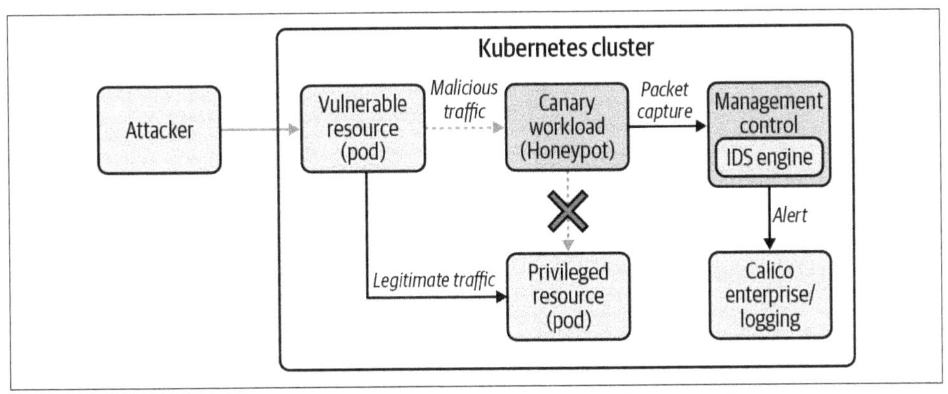

Figure 11-3. Sample implementation of honeypots in a Kubernetes cluster

Calico Enterprise has a honeypots feature that provides a supplementary detection method when strict network policies or monitoring is not feasible. Calico Enterprise honeypots work by deploying canary workloads and services in sensitive namespaces and monitoring for access. By leveraging Calico Enterprise's monitoring and alerting capabilities, any connections made to these canary workloads will generate an alert and can be traced back to the source. Canary traffic should be inspected using a DPI engine to provide signature-based detection to provide high-fidelity alerts and significantly reduce false positives.

DNS-Based Attacks and Defense

When you look at activity in a Kubernetes cluster, DNS is critical to your applications that are running. Kubernetes supports DNS as an infrastructure, and DNS support is available for using DNS names for pods and services. CoreDNS is the recommended DNS server for your Kubernetes cluster. Since DNS is critical to cluster operation, DNS traffic needs to be allowed inside the cluster and even for external lookups. This makes DNS an attractive option for adversaries to target. In this section we will cover domain generation algorithm (DGA) attacks that are used by adversaries to establish a connection to their command-and-control center and then for exfiltration of data.

Figure 11-4 shows how a domain generation attack works. The adversary first downloads an exploit inside the cluster that uses a known seed and an algorithm to generate domain names. The exploit then queries the algorithm-generated domain names. The same algorithm is run to spin up a DNS server that responds to DNS queries, and this cycle repeats till the client and the server domains match. Upon a successful match, the cluster has established a successful connection to the command-and-control server for the malware.

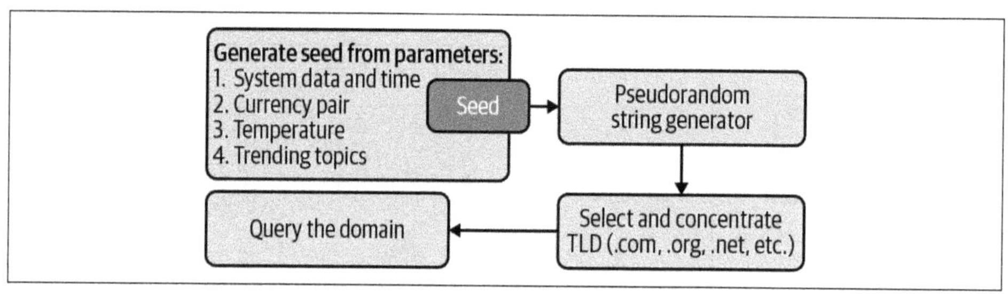

Figure 11-4. DGA-based attack

Since domain names are generated randomly using an algorithm and the queries are legitimate DNS queries, it is not possible to detect these types of attacks using DNS threat feeds or at the perimeter using DPI. Also, the fact that there is a relatively large amount of DNS activity in the cluster means it is easy for the malware to hide its activity inside the cluster. The way to detect these types of attacks is to use a machine learning technique that can predict a malicious domain just by analyzing the domain name. Another mechanism that can be effective is to use machine learning to baseline the number of DNS responses that do not resolve to a valid server and report an anomaly if there is an increase in such failed DNS queries.

You can implement a DGA detection mechanism by having the security research team collaborate with the data science team to build this mechanism. Calico Enterprise provides a DGA implementation integrated with its alerting engine.

Conclusion

In this chapter, we covered how you can implement threat defense in your Kubernetes cluster. The following are the key takeaways from the chapter:

- The techniques presented are based on our current research, and this area is constantly evolving, with adversaries using newer techniques and security teams working on solutions to counter these threats. We recommend your security team focus on threats seen, analyze them to evaluate if they are applicable to Kubernetes, and work on mitigation techniques.

- Kubernetes is a new technology, and we are starting to see it become a focus area for adversaries, so an effective threat defense strategy is required.
- It's very important to understand the cybersecurity kill chain and how it applies to Kubernetes in order to build an effective threat defense strategy.
- It is important for you to apply threat feeds and DPI-based techniques to traffic inside your cluster to detect attacks that originate inside the cluster. It is not adequate to rely on these techniques being applied only at the perimeter, as traffic originating inside the cluster may not traverse through devices at the perimeter.
- Honeypots and DGA-based attacks are examples of advanced threat defense techniques for your Kubernetes cluster that you should implement to thwart sophisticated attacks.

Conclusion

We hope that the book helped you understand how observability and security for Kubernetes deployments are different from traditional deployments. And we hope that the book is a guide for you as you design and implement your security and observability strategy, whether you are in the early stages of your journey or further along in adopting Kubernetes. The key takeaway is that you need to think about security and observability at every stage of your journey; it should not be an afterthought that is implemented once you have designed your deployment. We often hear folks say, "I will not need to worry about security or observability for a while; let me first get my workloads running in Kubernetes." This line of thinking is not right, as the right security implementation will likely alter the design and will likely cause an untimely iteration to the design and delay the implementation. The following are some of the characteristics that make Kubernetes different:

- Kubernetes is the most widely adopted orchestration engine for deploying modern applications and is used both in public cloud and on-premise deployments.
- Kubernetes is declarative in nature and enables users to specify outcomes for their application deployments (e.g., scale, specifications, access, etc.).
- Kubernetes continuously monitors the status of the deployment and takes corrective action to ensure the deployment is operating as specified.
- Kubernetes abstracts the details of networking, IP addresses, etc., and instead allows users to define identity using higher-level constructs like labels.
- Due to these characteristics, implementing observability and security for Kubernetes needs a different approach.

These characteristics of Kubernetes have also altered the development process and the teams involved. Previously, development teams built applications that would be deployed on preprovisioned infrastructure. In today's world, when you deploy an application in the cloud, you first provision the infrastructure (e.g., hosts, VMs, etc.)

that is required for your application and then deploy your application on the provisioned infrastructure. Also, the infrastructure is dynamic and adapts to the needs of the application. The following lists the life cycle of an application in a Kubernetes cluster and the role of various teams:

- The deployment of an application comprises the build stage (create the resources needed for the application), the deploy stage (deploy the application using Kubernetes), and the runtime stage (application operation post deployment).
- The teams that are responsible for successful deployment and operation of the application are the operations team, the platform team, the networking team, the security team, and the compliance team.
- In order to design an effective security and observability strategy for your application, you need to consider security and observability at all stages.
- Collaboration between the teams is critical to success as security is a joint responsibility of all the teams involved.

The following checklist is a good guide to ensure that you have an effective security and observability implementation for your Kubernetes deployments:

- All images are scanned for known vulnerabilities prior to deployment and then periodically scanned for vulnerabilities discovered post-deployment.
- All container images being deployed are built with minimal base OS components (e.g., distroless or scratch images).
- The operating system on the host is an immutable Linux distribution that reduces the attack surface area.
- The host OS and the pods deployed are configured with controls that only allow required access (e.g., system calls, filesystem access).
- The Kubernetes cluster deployment is hardened with the encryption of secrets, securing access to the API server and the data store.
- Deployment of workloads in your Kubernetes cluster is controlled by best practices for RBAC and admission controllers to enforce policies.
- All access to services in your cluster is exposed to external clients using security best practices.
- The Kubernetes cluster has an integration with a perimeter security device (firewall or a gateway) to enable the device to have visibility into traffic originating from the cluster so it can add effective controls.
- You need to ensure that access control is in place for network traffic and application traffic using L3/L7 policies.

- Ensure that you are using a tool native to Kubernetes to implement observability; e.g., your tool needs to be aware that pods backing a service or a deployment are identical and should be viewed as a unit to "observe" a service.
- Ensure that you implement machine learning to baseline behavior of various entities in your cluster and build an anomaly detection system layered above that to detect security incidents.
- Ensure that you implement threat defense features like IDS, IPS, and advanced threat detection techniques in your cluster to detect malicious activity inside your cluster.
- Ensure that you have implemented data in transit encryption for communication inside the cluster as well as communication to external entities.

We wish you the best in your Kubernetes journey as you implement security and observability for your workloads!

Index

A

B

C

D

E

F

G

H

I

J

K

N

O

P

T

U

V

W

X

Z

About the Authors

Brendan Creane is head of engineering at Tigera, where he's responsible for all engineering operations, including Calico Cloud, Calico Enterprise, and Project Calico. Brendan has several decades of experience building enterprise security, observability, and networking products.

Amit Gupta is vice president of product management and business development at Tigera, where he's responsible for the strategy and vision of Tigera's products and leads the delivery of the company's roadmap. Amit is a hands-on product executive with expertise in building software products and services across various domains, including cloud security, cloud native applications, and public and private cloud infrastructure.

Colophon

The animal on the cover of *Kubernetes Security and Observability* is an African fish eagle (*Haliaeetus vocifer*), a large species of eagle found throughout sub-Saharan Africa. Its scientific name means "vociferous sea eagle."

Female African fish eagles are about 7 pounds and have an almost 8-foot wingspan. Males are somewhat smaller at 4.5–5.5 pounds, with a 6.5-foot wingspan. This sexual dimorphism is typical in birds of prey. Adults have a mostly brown body, with a white head, chest, and tail. The wings are black.

In their range, the eagles are found near freshwater lakes, resevoirs, and rivers. Sea eagles have spiricules on their toes; these structures allow the eagle to grasp fish and other slippery prey, such as waterbirds. It can carry off prey many times its own size, either by dragging the prey across the surface of the water or by dropping into the water and paddling to shore with its wings.

The African sea eagle is the national bird of Namibia and Zambia. It appears on the coat-of-arms and on the flags of several African nations. Its conservation status is Least Concern. Many of the animals on O'Reilly covers are endangered; all of them are important to the world.

The cover illustration is by Karen Montgomery, based on a black and white engraving from *Lydekker's Royal Natural History*. The cover fonts are Gilroy Semibold and Guardian Sans. The text font is Adobe Minion Pro; the heading font is Adobe Myriad Condensed; and the code font is Dalton Maag's Ubuntu Mono.

CPSIA information can be obtained
at www.ICGtesting.com
Printed in the USA
LVHW112309191121
703844LV00007B/501

9 781098 107109